2011 AUBE

前言
PREFACE

渐渐地顿悟

西藏，拉萨，大昭寺前，晨曦里，香火中，一汉子向着大殿，下跪、扑身、五体投地、起身，一气呵成。他在与佛祖对话？在与自己对话？这是宗教吗？我迷惑。

法国，朗香教堂，2003年一场大雪后，一位日本建筑学学生站在雪地里，顺着雪一样白的墙面，仰头望向屋顶。他在寻找？在圆梦？

西班牙，毕尔巴鄂，2005年春，我按图索骥，寻找改变了这个城市的那座建筑，"这条街右转直走就到了"，转向，一瞬间我看到了那个天外来客……我究竟在找什么？

深圳，某写字楼会议室，2006年夏，业主咆哮，问责，攻击，泄愤。是我们不负责？不用心？不努力？还是我们在坚持什么？

深圳，某餐厅，同学聚会中，"某某当年不怎样吧，人家现在……""你早该出来自己做……"

……我问我自己，他们说的是我想要的？不止吧？

深圳，生态广场C栋，2011年某晚，众人加班，刚完成了一张草图的我点支烟，侧头看看自己的团队，他们为项目？为家？为什么？

西藏，拉萨，大昭寺前，八廓街上，我又看到了那个汉子，众人中面目祥和，眼神安定坚毅。忽然想起一位朋友的话：执着于自己的梦想，就是信仰。执着于自己的信仰，就是宗教。

欧博设计

设计董事　杨光伟

THE JOURNEY TO INSIGHT

Jokhang Temple, Lhasa, Tibet. It was a dawn with burning incense. Facing the main hall, a man kneeled down, prostrated and got up at one stretch. Was he conversing with Buddhist or himself? Was that religion? I was confused.

Chapelle de Ronchamp, France. It was in 2003 and had just snowed heavily. A Japanese architecture major stood in snow and looked over snow-white walls and up the roof. Was he looking for something? Was he interpreting a dream?

Bilbao, Spain. It was in the spring of 2005. Following a map, I tried to find the building that had changed the city. I walked along the street and turned right, as I was told. And the heavenly building came right into sight. What on earth was I looking for?

A conference room of an office building in Shenzhen. It was in the summer of 2006. Proprietors were shouting, complaining, blaming and venting their resentment. Were we irresponsible? Did we not try our best? Or were we just persisting in something?

A restaurant, Shenzhen. It was a classmate reunion. "That guy seemed not promising back then, but now ……" "You should have become self-employed long ago……"……I asked myself whether what they talked about was what I wanted? Or was there more?

Building C of the Ecological Plaza, Shenzhen. It was a night in 2001 and employees were overworking. Having finished a draft, I lighted a cigarette and then looked at my team. Were they working so hard for projects, for their family or for something else?

Jokhang Temple, Lhasa, Tibet. On the Barkhor Street, I met the man once again. He was calm, firm and peaceful among crowds, which reminded me of what one friend had told me----faith is to persist in one's dream and religion is to persist in one's faith.

AUBE Conception
Design Director : Yang Guangwei

目录
CONTENTS

ZHONGTIAN · FUTURE ARK, GUIYANG 贵阳中天·未来方舟

贵阳中天·未来方舟A区文化活动中心建筑及景观设计
Architectural and Landscape Design of The Culture Center of Zhongtian · Future Ark, Guiyang 008

贵阳中天·未来方舟样板房建筑设计
Architectural Design of the Show Flat of Zhongtian · Future Ark, Guiyang 018

贵阳中天·未来方舟F3地块建筑设计
Architectural Design of The Lot F3 of Zhongtian · Future Ark, Guiyang 024

贵阳中天·未来方舟-东懋中心建筑设计
Architectural Design of the Dongmao Center of Zhongtian · Future Ark, Guiyang 032

URBAN PLANNING 规划

三亚万科湖心岛修建性详细规划及建筑方案设计
Detailed Planning and Architectural Design of Vanke Center Island Project in Sanya 040

深圳盐田翡翠岛项目规划、建筑及景观设计
Planning, Architectural and Landscape Design of the Feicui/Jade Island, Yantian, Shenzhen 048

唐山湾国际游艇社区项目概念方案规划设计
Planning Design of the International Yacht Community of Tangshan Wan 054

苏州太湖国家旅游度假区东入口地区城市设计
Urban Design of the East Entry of Taihu Lake International Tourism Zone, Suzhou 060

深圳湾生态科技城B-TEC项目规划及建筑方案设计
Planning&Architectural Design of B-TEC City of Shenzhen Bay 068

长城笋岗城市综合体规划及建筑概念方案设计
Planning&Architectural Design of the Great-wall Complex Building In Sungang, Shenzhen 084

浙江省余杭高新技术产业园区科技园一期工程规划及建筑设计
Planning&Architectural Design of the Yuhang High-tech Industry Park (phase I), Zhejiang 088

杭州转塘(1-C2-03地块)项目规划及建筑设计
Planning&Architectural Design of the Lot 1-C2-C3 of Zhuantang Area, Hangzhou 094

ARCHITECTURE 建筑

深圳市塘朗车辆段A区建筑概念方案设计
Architectural Design of the Parcel A of Tanglang Rolling Stock Depot, Shenzhen 102

南海会馆概念性建筑设计
Architectural Design of The South Sea Memorial Hall, Foshan 106

百度国际大厦建筑设计
Architectural Design of Baidu International Tower 110

深圳市侨城北工业区升级改造工程建筑设计
Architectural Design of the Upgrading & Renewal Project of the Industrial Park in the North of Oct, Shenzhen **116**

城建观澜仁山智水花园天际会所建筑设计
Architectural Design of The Horizon Club of Expander · The Garden of Benevolence & Wisdom, Shenzhen **122**

中铁南方总部大厦建筑设计
Architectural Design of the Crsid Tower **126**

昆明滇池湖岸花园项目建筑设计
Architectural Design of the Dianchi Lakeside Residence, Kunming **130**

国家（青岛）通信产业园1、2、3号地块规划、建筑、景观方案设计
Planning, Architectural & Landscape Design of the Plots 1/2/3 of the Communication Industry Park of Qingdao **136**

阿里巴巴深圳大厦建筑概念方案设计
Architectural Design of Alibaba's Branch Tower in Shenzhen **140**

深圳市宝安区新安宝城34区改造项目建筑设计
Architectural Design of the Renewal Project of the 34th Block of Bao'an District, Shenzhen **146**

深圳职业技术学院西校区综合楼工程建筑设计
Architectural Design of the Multiple-use Building in the West Campus of Shenzhen Polytechnic **150**

LANDSCAPE 景观

无锡地铁1号线胜利门站广场景观方案征集
Landscape Design of the Victory Gate Station Square of Wuxi Metro Line 1 .... **158**

无锡地铁1号线出入口、风亭外观景观效果及下沉广场景观方案征集
Landscape Design of The Exit of Wuxi Metro Line 1, Its Wind Pavilion and Sunken Square **168**

贵阳云岩渔安安井回迁安置居住区D组团景观设计
Landscape Design of the Group D of the Relocation Residential Area in Anjin, Yuan, Yunyan, Guiyang **178**

贵阳规划展览馆景观设计
Landscape Design of Guiyang Planning Exhibition Hall **186**

昆明滇池湖岸花园桂园中心花园
Landscape Design of the Osmanthus Central Garden of the Dianchi Lakeside Residence, Kunming **190**

深圳南海中学景观方案设计
Landscape Design of Nanhai (South Sea) Middle School, Shenzhen **196**

贵阳民族盛会广场景观方案设计
Landscape Design of the National Minorities Festival Square in Guiyang **200**

AUGE2011

AUBE2011

贵阳中天·未来方舟
ZHONGTIAN · FUTURE ARK, GUIYANG

ARCHITECTURAL
AND LANDSCAPE DESIGN OF THE CULTURE CENTER OF ZHONGTIAN · FUTURE ARK, GUIYANG
贵阳中天·未来方舟A区文化活动中心建筑及景观设计

ARCHITECTURAL DESIGN 建筑设计

Under the enlightment of Zhongtian's corporate culture that the present is the future and taking the basic funtions of a sales center, this project is meant to introduce and promote Yu'an to clients. Organically, the culture center is divided into three sections which form the following sequence (cloud box-rock box-magic box) in the transition from the present to the future. The transparent shell of the culture center submerged in mist and cloud, As a time tunnel supported by independent structure, cloud offers visitors with various sensory temptations and leads them to the second part of the sequence - the rock box, where they experience astonishing high technology. The whole visit will reach its peak at global screens. In the end, magic box, which contains the near future, shows visitors Zhongtian's corporate image and the beautiful future home.

"现在就是未来"是中天的企业文化。受其启发并结合销售中心的基本诉求，项目向客户展示渔安并达到推广之目的。文化活动中心有机分散为三个体量，形成了由"现实向未来"过渡的三个序列：cloud box-rock box-magic box。 透明的外壳袅绕着轻舞的云雾，包裹着若隐若现的cloud，cloud作为独立结构支撑的时光隧道，让游客经历过各种感官的诱惑之后，进入rock box来到第二个建筑序列，这里包含了震撼的高科技体验，并在环球影幕处达到游艺的高潮。最后，magic box 销售中心，容纳了触手可及的未来，向游客展示着中天的企业形象，未来的美好家园。

LANDSCAPE DESIGN 景观设计

First, the project carries forward the building's design principle, strengthening its design concept and establishing the relations between the building and its surrounding water.

The soil, dug from mountains in the west of the culture center, will be used to form a hill on its south.The water will be led to the culture center to forme a mirror-like pool. In this way, water systems in the culture center will have a visual and conceptual source.

Second, establishing relations between the building and land

Landscape water systems will be connected through such landscaping factors as cascades and lake views so as to form a recyclable whole.

Third, ecology and sustainability

Native plants are used to create wetland landscape and strengthen water's self-purification. And a great importance is attached to the original ecological landscape and environmental sustainability.

延续建筑的设计原理、强化建筑设计概念，并建立联系：

延续建筑被水包围的宏伟气势。从文化活动中心东侧山体挖出的土方量填充至文化活动中心南侧塑造山体，并引水至文化活动中心形成镜面水体，为其水系从视觉和概念上找到源头。

建立建筑、地块之间的联系：

通过叠水、湖景等景观要素使景观水系相互连接和流动，文化活动中心水系成为一个可循环的整体系统。

生态及可持续性：

运用原生植物，营造湿地景观，强化水体自净功能，注重原生态景观和环境持续性发展。

Client : Guiyang Real Estate Development Co. Ltd. of Zhongtian Urban Development Group
Location : Yunyan District, Guizhou
Land area : 7.2ha
Building area : 21 000 m²
Landscape area : 60 000 m²
Height : 21.2m
Function : sales center
Under construction

客　户：中天城投集团股份有限公司
位　置：贵阳市云岩区
用地面积：7.2hm²
建筑面积：2.1万m²
景观面积：6万m²
建筑高度：21.2m
主要功能：销售中心
在建项目

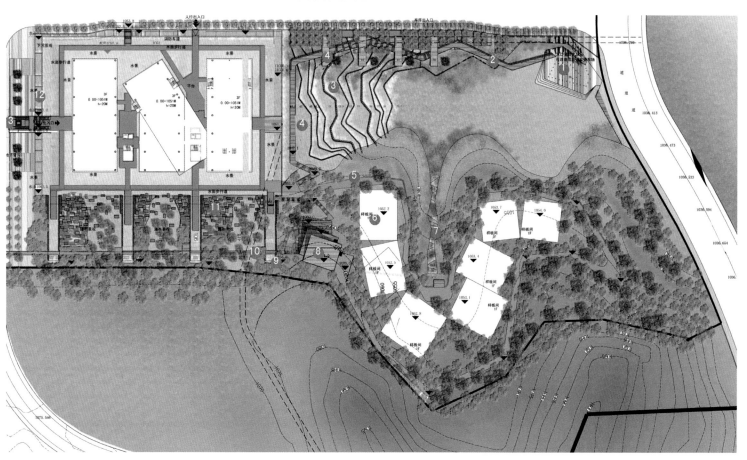

ARCHITECTURAL DESIGN 建筑设计

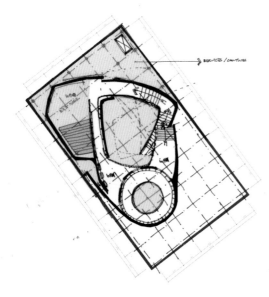

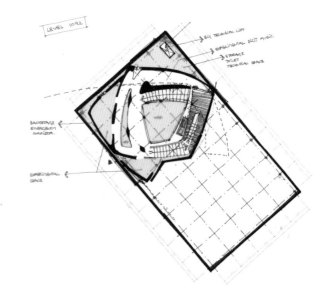

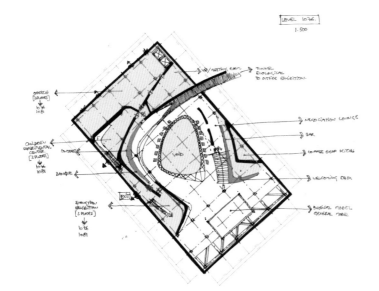

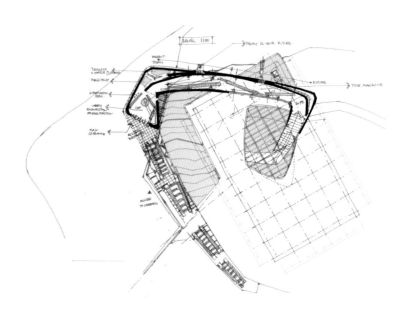

夜景效果图

LANDSCAPE DESIGN 景观设计

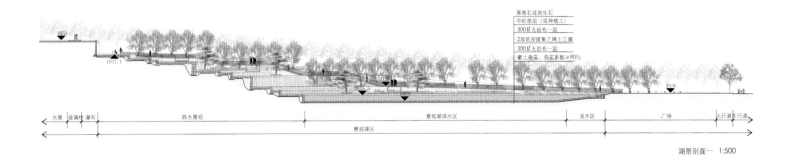

湖景剖面一 1:500

湖景剖面二 1:250

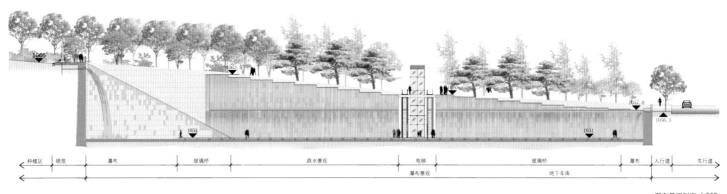

瀑布景观剖面 1:350

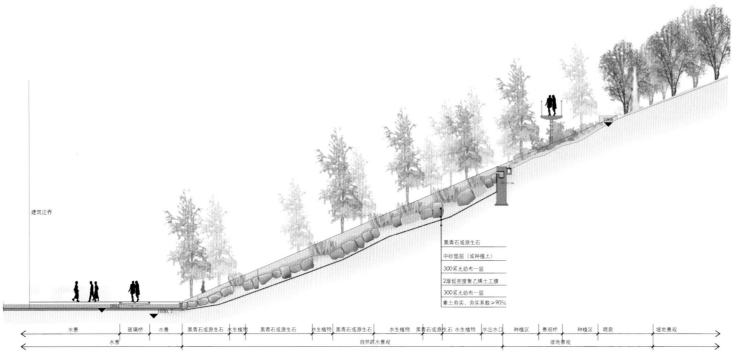

原生石跌水剖面 1:150

建成实景图

ARCHITECTURAL
DESIGN OF THE SHOW FLAT OF ZHONGTIAN · FUTURE ARK, GUIYANG
贵阳中天·未来方舟样板房建筑设计

Client : Zhongtian Urban Development Group Co.,Ltd.
Location : Yu'an District, Guiyang
Land area : 441m^2
Height : <9m
Function : show flat
Constructed on 2011

客　　户：中天城投集团股份有限公司
位　　置：贵阳渔安片区
用地面积：441m^2
建筑高度：<9m
主要功能：样板房单元展示
2011年已建成

Situated on the slop in the southeastern side of Yu'an Museum, this project demonstrates some residential units in the project of Future Arc and also serves as a small exhibition hall.

Taking local traditional elements and images as the design core of every unit, the design endeavors to form connections between the future image of the project and the contemporary, even past culture in Guiyang. Means to do so include using traditional materials with symbolic meanings like bamboos and vines as buildings' envelopes, incorporating traditional hanging houses in the design of basic images and so on, which all indicate respect and reference to local tradition.

本项目位于渔安博物馆东南面的山坡上，选取未来方舟项目中的部分居住单元在此展示，同时具备小型博览会的功能。

设计着重于发掘贵州当地的传统元素及意象表达，作为每个单元体的主导设计核心，为项目的未来意象与贵州当代乃至过去的文化传承之间创造交点。手段包括采用具有象征意义的传统材料—竹或藤作为建筑的表皮及外围护结构，或是借用传统吊脚楼的形制为基础意象进行设计等，均表达了对当地传统的尊重和借鉴。

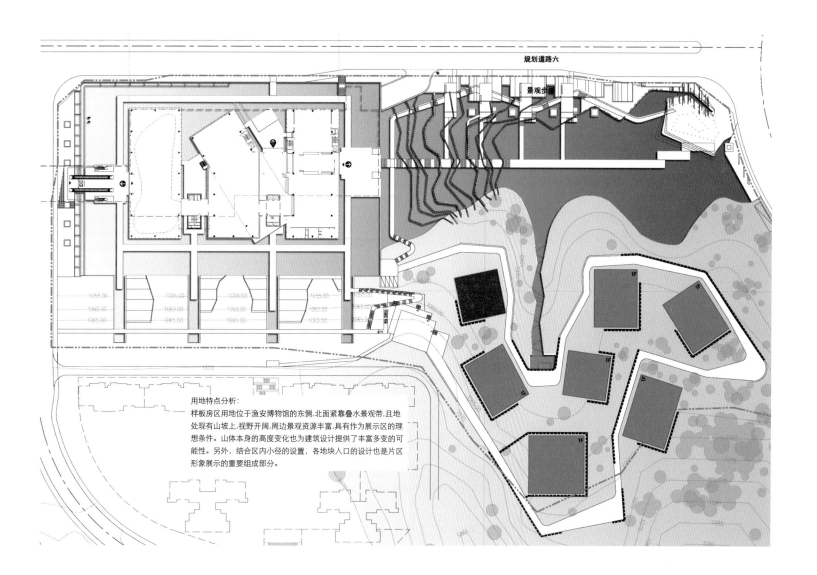

SCHEME 1 方案一

021 · ZHONGTIAN · FUTURE ARK, GUIYANG

贵阳简称"筑",盛产竹,康熙二十六年,立贵阳府设贵筑县,即今之贵阳。

现今贵阳市中心最繁华的大十字,唐宋年间称"黑羊菁","菁",亦即竹林,因此贵阳和竹子深有渊源。

本方案采用贵州盛产的竹子为主要建造材料,环保并具有浓厚的地方特色。采用竹子,造型可丰富多变并且造价低廉,为理想的建筑材料。

并且建筑形制由吊脚楼演化而来,尊重传统并有所创新,架空的回廊可让参观者从多角度欣赏建筑,如展示艺术品一般。另外,自然的材料与项目所处的山坡能很好地融为一体。

SCHEME 2 方案二

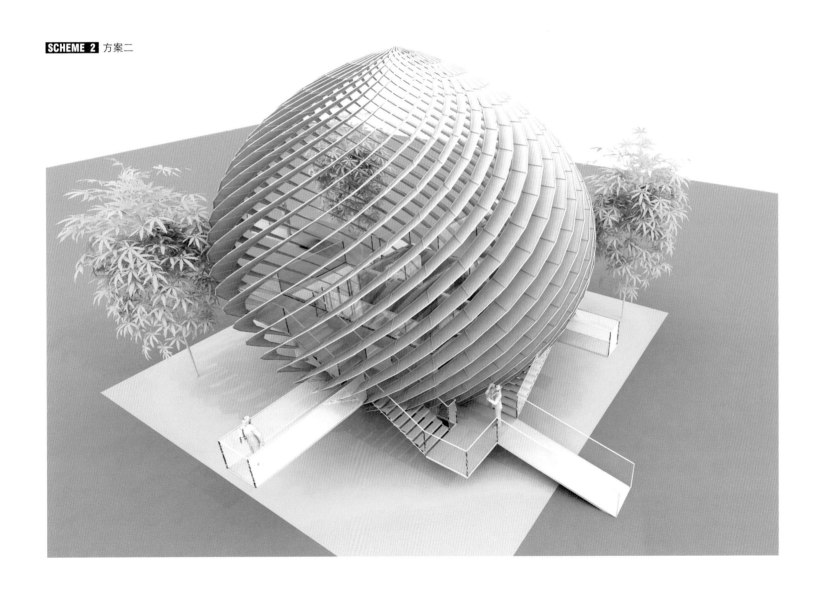

023 · ZHONGTIAN · FUTURE ARK, GUIYANG

SCHEME 3 方案三

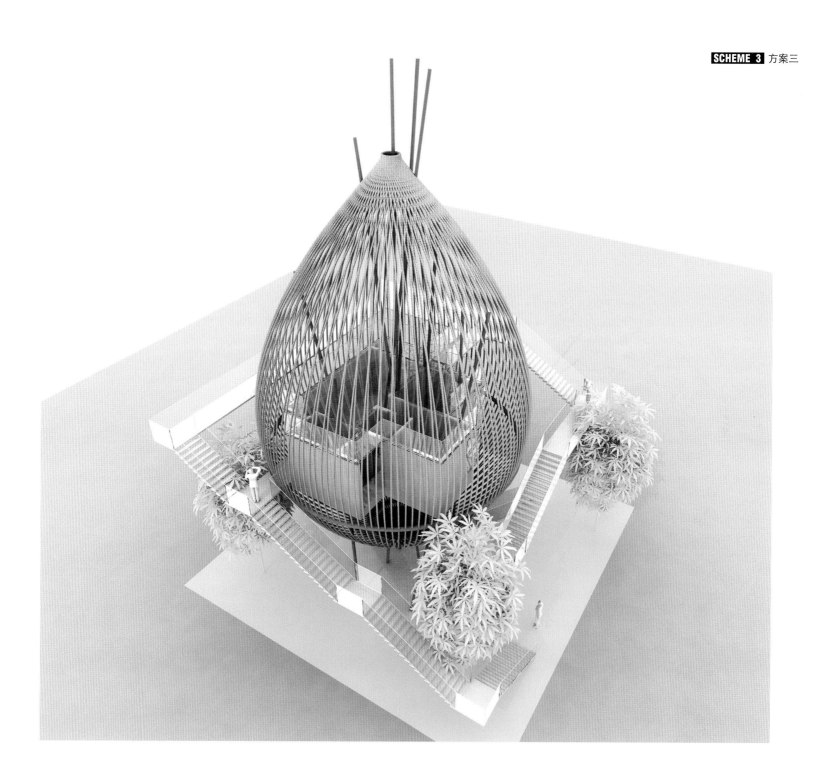

ARCHITECTURAL
DESIGN OF THE LOT F3 OF ZHONGTIAN · FUTURE ARK, GUIYANG
贵阳中天·未来方舟F3地块建筑设计

Client : Zhongtian Urban Development Group Co.,Ltd.
Location : Yu'an District, Guiyang
Land area : 2.98ha
FAR : 4.5
Building area : 169 000m²
Height : 280m
Function : office, hotel, apartment, retail

客　　户：中天城投集团股份有限公司
位　　置：贵阳渔安片区
用地面积：2.98hm²
容 积 率：4.5
建筑面积：16.9万m²
建筑高度：280m
主要功能：办公、酒店、公寓、商业

Time Gate

As this project occupies an important position, standing between the old downtown and Yu'an District, the design combines all functions into a simple yet magnificent single whole that looks like a giant gate linking the past and the future. It indicates that the project pays much attention to historical heritages while looks into the future. Giant drawbridge-like overhangs further interpret the theme of connection and transition and emphasize the boundaries of buildings.

时光之门
本案处于从旧城区进入渔安片区的门户区，位置显要，方案将所有功能合并为简洁大气的单一整体，形体意象如一道巨门，连接过去和未来，体现项目对历史的延续传承以及对未来的展望，彰显出项目的门户身份。
巨型结构的悬挑，暗示了吊桥形象，进一步阐述了"联系"与"穿越"这一主题，强化了建筑本身的边界意识。

SCHEME 1 方案一

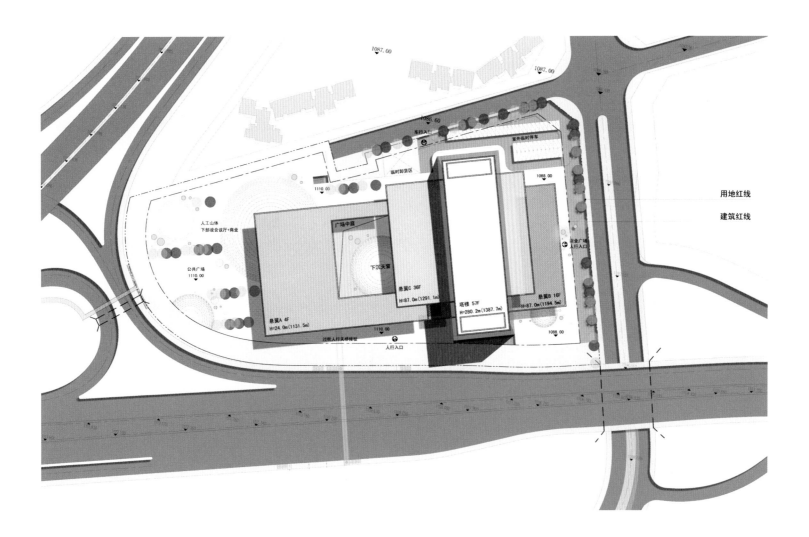

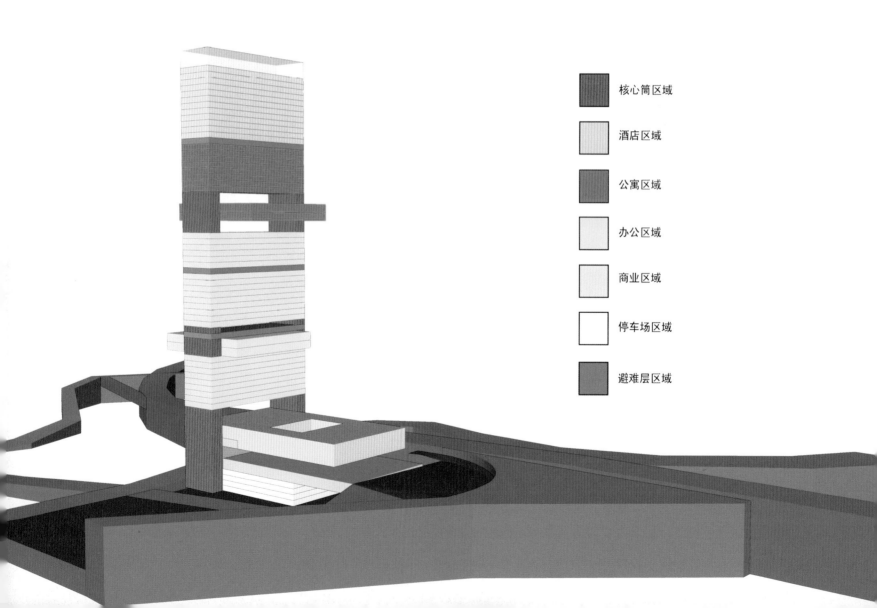

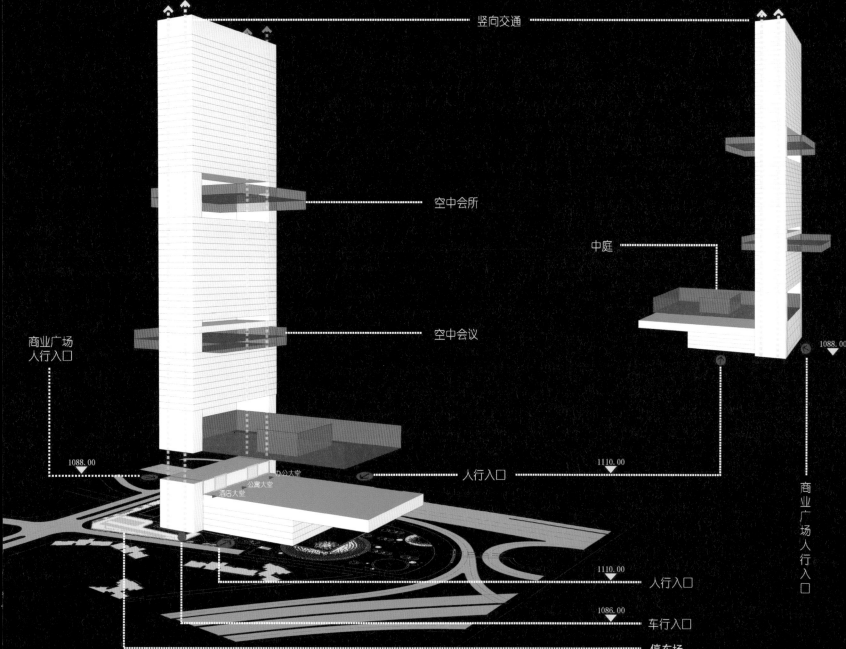

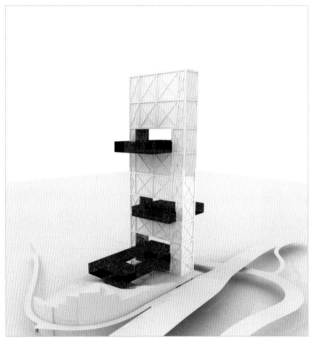

029 · ZHONGTIAN · FUTURE ARK, GUIYANG

SCHEME 2 方案二

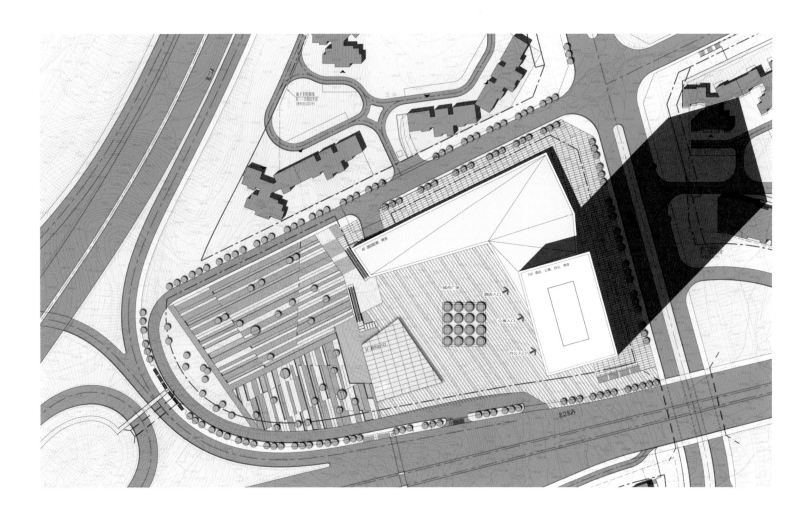

ARCHITECTURAL DESIGN OF THE DONGMAO CENTER OF ZHONGTIAN · FUTURE ARK, GUIYANG

贵阳中天 · 未来方舟—东懋中心建筑设计

Client : Zhongtian Urban Development Group Co.,Ltd.
Location : Yu'an District, Guiyang
Land area : 9.84ha
FAR : 6.78
Building area : 667 400m²
Height : 350m
Function : Office, hotel, apartments, retail

客　　户：中天城投集团股份有限公司
位　　置：贵阳市西部渔安片区
用地面积：9.84hm²
容 积 率：6.78
建筑面积：66.74万m²
建筑高度：350m
主要功能：办公、酒店、住宅、商业

Located in the south of the project of Future Ark, Dongmao Center is its core business zone and also its southeastern portal. Highlighting the concept of portal, the design attempts to construct a modern high-end commercial center with the Chinese concept of portal, using four dragon pillars.

东懋中心位于未来方舟整体项目南部，是该项目的核心商务区，同时也是项目的东南门户。
设计中突出"门户"的概念，用四根盘龙柱的造型打造出具有中国门户概念的现代高端商务中心。

033 · ZHONGTIAN · FUTURE ARK, GUIYANG

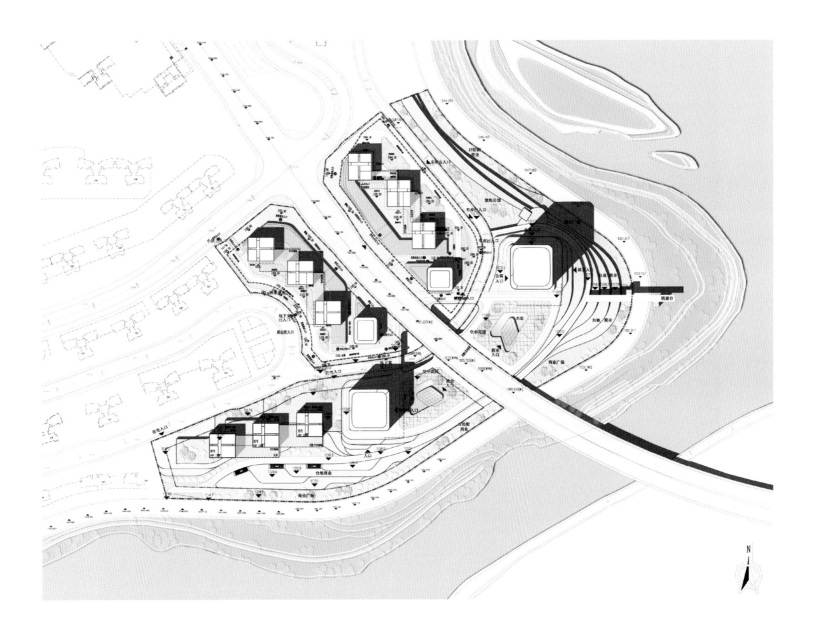

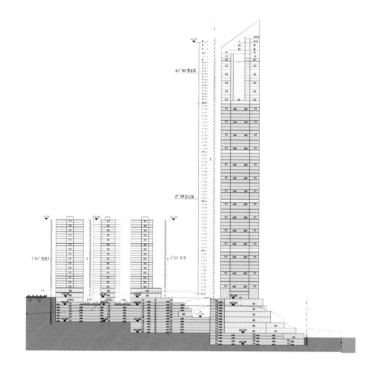
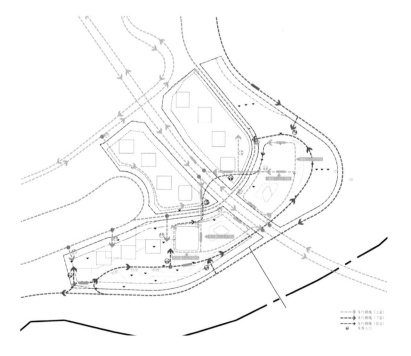
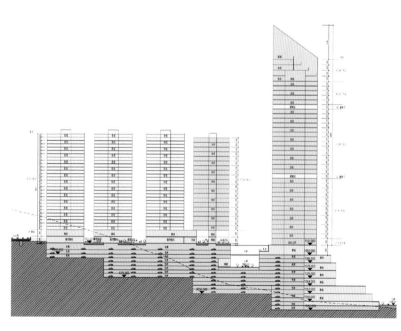

037 · ZHONGTIAN · FUTURE ARK, GUIYANG

AU8ES011

AUBE2011

DETAILED PLANNING
AND ARCHITECTURAL DESIGN OF VANKE CENTER ISLAND PROJECT IN SANYA
三亚万科湖心岛修建性详细规划及建筑方案设计

Client : Hainan Vanke Real Estate Co., Ltd
Location : Sanya Hainan
Land area : 43.61ha
FAR : B-06-04:1.19,
　　　B-05-03:1.40,
　　　B-05-09:0.34
Building area : 450 000m²
Height : 50m
Function : football training, retail, apartments
International bidding : winning project

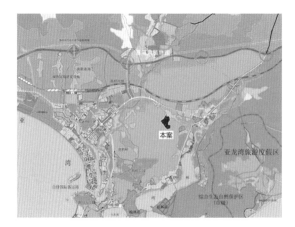

客　　户：海南万科房地产有限公司
位　　置：海南三亚市
用地面积：43.61hm²
容 积 率：B-06-04:1.19
　　　　　B-05-03:1.40
　　　　　B-05-09:0.34
建筑面积：45万m²
建筑高度：50m
主要功能：足球训练基地、商业、住宅
国际竞标：中标方案

With the central square as the spatial kernel, infrastructures such as hotels, retail, apartments, school and parks are organized in an effective way. And the central square can be used for daily activities like leisure activities, sports, entertainment and performances.

The planned structure of the project can be summarized as one center, one loop and three axes. One center refers to the central square.

One loop refers to the cycling loops.

Three axes refer to the three significant spatial axes formed through the combination of such factors as the planning of this area, its natural geographical conditions and its ecological landscape.

利用中心广场形成空间核心，将酒店、商业、居住、学校、公园等不同功能有效地组织在一起。同时该广场作为行为核心，还兼顾了人们休闲、运动、娱乐、表演等日常活动。

项目的规划结构可以概括为："一心、一环、三轴线"。

"一心"是指一个"空间中心"。

"一环"是指一个"自行车赛道环线"。

"三轴线"是指通过场地的规划条件、自然地理条件及生态景观要素，形成的具有相关意义的三条"空间轴线"。

CONCEPT 规划理念

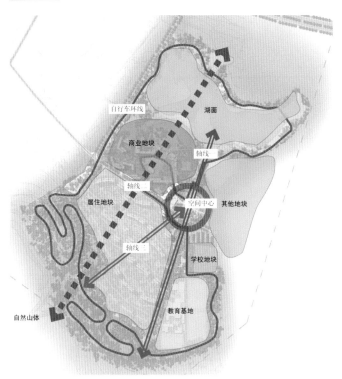

PEDESTRIAN SYSTEM 步行系统规划图

GONERAL INTRODUCTION 规划结构

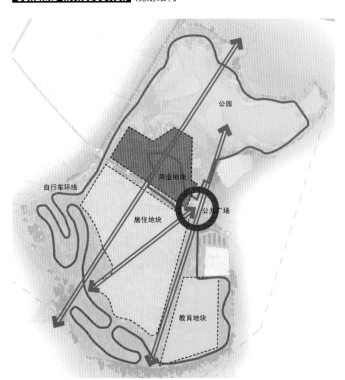

LANDSCAPE PLANNING 景观系统规划图

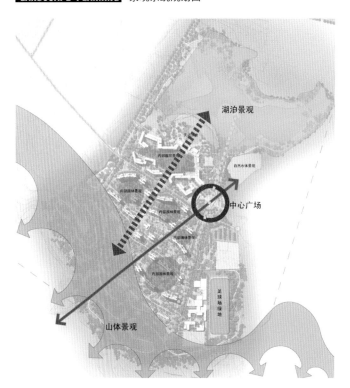

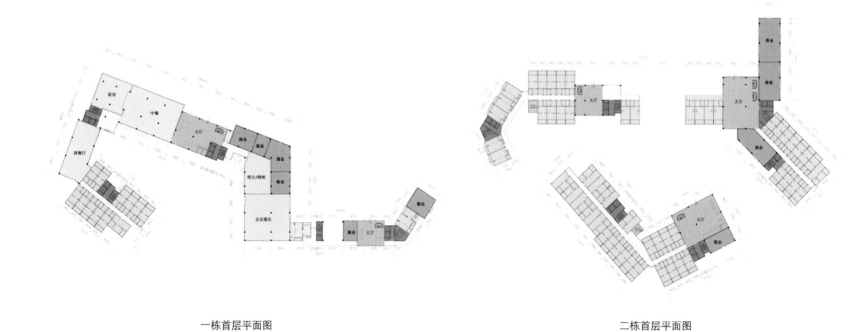

一栋首层平面图

二栋首层平面图

三栋首层平面图

走廊
Corridor

核心筒
Vertical Circulations

酒店
Appartments

酒店大堂及服务
Lobby and services

酒店服务间
Hotel services

商业
Shop

045 · URBAN PLANNING

PLANNING,
ARCHITECTURAL AND LANDSCAPE DESIGN OF THE FEICUI/JADE ISLAND, YANTIAN, SHENZHEN
深圳盐田翡翠岛项目规划、建筑及景观设计

Client : Shenzhen Yantian Port Group Co., Ltd.
Location : Yantian District , Shenzhen
Land area : 5.5ha
FAR : 2.36
Building area : 130 000m²
Height : 159.5m
Function : Hotel, office, retail
International bidding : 3rd prize

客　　户：深圳盐田港集团有限公司
位　　置：深圳市盐田区
用地面积：5.5hm²
容 积 率：2.36
建筑面积：13万m²
建筑高度：159.5m
主要功能：酒店、办公、商业
国际竞标：第三名

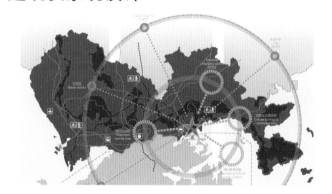

Located at the intersection of hills, rivers and sea, between a port and a tourist city, and at the junction of fast main roads, greenways and coastal plank roads along cliffs, the area where the project lies is an epitome of the spatial characters and the unique culture of the Yantian District and the Yantian Port.

In the planning of the project, the land among the Yantian High-speed, the Pearl Avenue and the Yantian Port is regarded as a whole centering on the refuge harbor. Walking paths and green strips on the land form a complete network, making the waterside land much more pleasant. The whole land will be the future activity center of Yantian, and the Feicui/jade Project will be the core of the center.

用地所在片区，是山、河、海的交汇处，港口和旅游城市的融合处，快速交通干线和绿道、滨海栈道的衔接处。她集中体现了盐田区和盐田港的空间特征和文化精神。

规划中将盐田高速、明珠大道、盐田港之间的片区作为一个整体，整个片区围绕着避风港水域组织在一起。片区内步行和绿化形成完整的网格，打造亲水、宜人的片区氛围。整个片区将形成未来盐田的活动中心，而翡翠岛项目就是这个中心的核心。

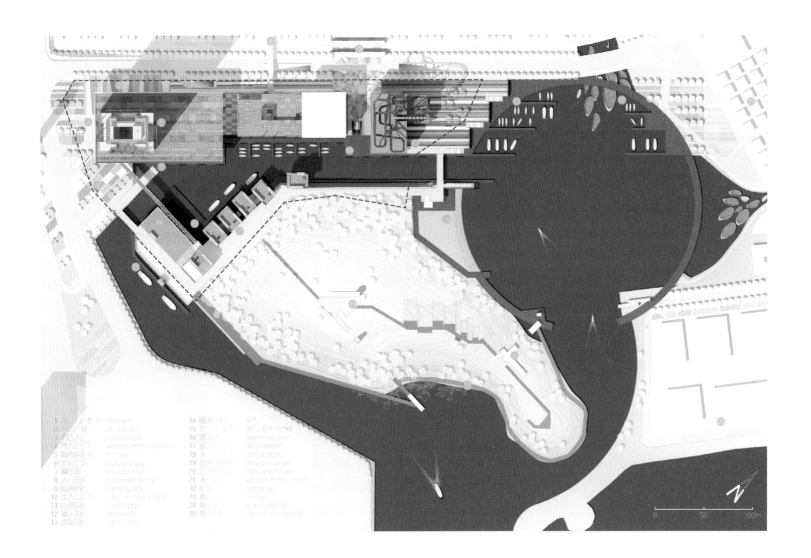

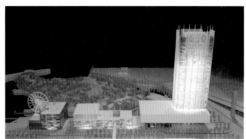

PLANNING
DESIGN OF THE INTERNATIONAL YACHT COMMUNITY OF TANGSHAN WAN
唐山湾国际游艇社区项目概念方案规划设计

Client : Tangshanwan International Yachting Development Co., LTD
Location : The outlet to the Bohai Sea, Tangshan
Height : 60m
Function : detached house and yacht club
International bidding : winning project

Real estate development zone :
Land area : 82.2ha
FAR : 0.69
Building area : 568 000m^2

Non-real estate development zone :
Land area : 75.7ha
FAR : 0.06
Building area : 46 000m^2

客　　户：唐山湾国际游艇发展有限公司
位　　置：唐山市沿渤海湾河道出海口
建筑高度：60m
主要功能：独栋住宅及游艇会所
国际竞标：中标方案

房地产开发部分：
用地面积：82.2hm^2
容 积 率：0.69
建筑面积：56.8万m^2

非房地产开发部分：
用地面积：75.7hm^2
容 积 率：0.06
建筑面积：4.6万m^2

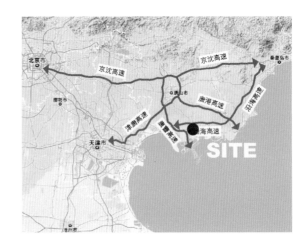

The Project of International Yachting Community in Tangshanwan is a high-end real estate project on the theme of yacht. The planning concept is to connect the northern sea with the southern sea so that the layout of the community will be in line with the geographical feature of the new tourist center.

This layout featuring an axis that runs from northern sea to southern sea highlights its leading role in the project, emphasizes the specialties of the coastal community, and strengthens the tie between each part of the project and the coastal area.

The symmetrical layout manifests the grand brand image of the project. Meanwhile, it matches well with the mirror reflection of the ocean and contains the cultural connotation of water.

唐山湾国际游艇社区项目是一个围绕游艇为主题展开的高端房地产项目。规划理念是尽量打通基地南北向与海的联系，使规划布局与新旅游中心的城市肌理相吻合。
这种强调贯穿南北，通向海洋的轴线的规划布局特点，凸显了其在项目中的龙头作用；
凸显了滨海社区的特性，加强了项目各区域至滨海的通达性。
规则以对称的平面布局，彰显出项目恢弘大气的品牌形象，同时与海洋、水体的"镜面反射"相呼应，蕴含了"水"的文化内涵。

CONCEPT 规划设计概念

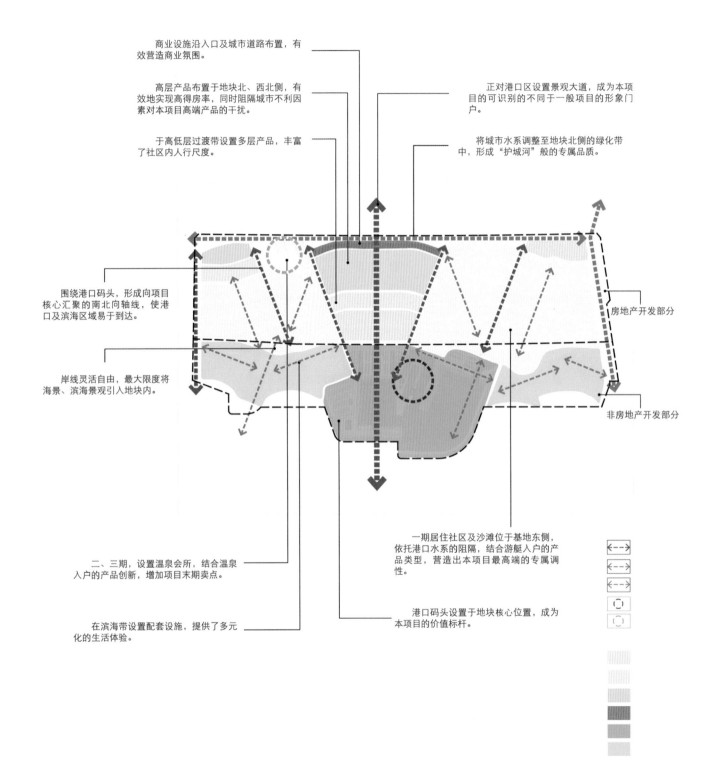

CONCEPT 规划设计概念

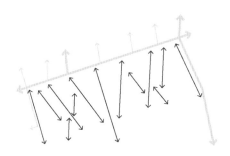

根据项目用地的朝向和所处环境,我们的规划理念是尽量打通基地南北向与海的联系,使规划布局与新旅游中心的城市机理相吻合。

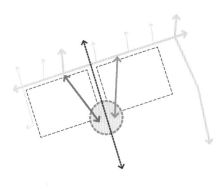

凸显了其在项目中的龙头作用,形成以港口码头为核心并易于到达的规划格局,并作为各分期开发的价值标杆;凸显滨海社区的特性,加强项目各区域至滨海的通达性。以规则对称的平面布局,彰显项目恢弘大气的品牌形象,同时与海洋、水体的"镜面反射"相呼应,蕴含了"水"的文化内涵。

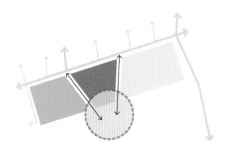

各分期的主要景观轴线均以港口码头为指向,清晰可行的划分出各开发区。

THEME AND PRINCIPLE 设计主题及原则

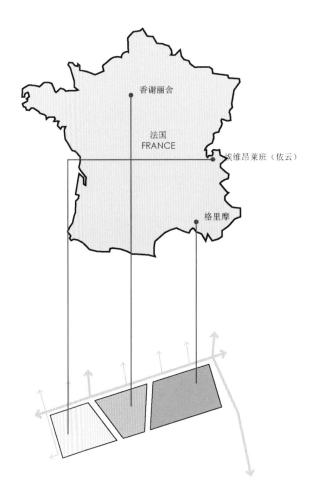

在景观和建筑的设计上采用的主题风格:格里摩(游艇别墅区)、香舍、依云桥舍三大区的特色,打造贵族级住宅的生活模式和典范。

游艇码头居住社区,私人游艇可通达至各户花园,是本项目的个性形象标识。我们为其植入了第一次由世界知名游艇大师Spoorry在创建法国南部格里摩时所提出的设计主题,并作为设计灵感的起源。

这个概念利用引入水系景观方法来打造项目中开发的地块,使建筑与水环境互动,并使光线、景色在水面波动中产生变化,形成美丽而诱人的独特景致。

"温泉"是为二期,三期增设的亮点,我们引入了著名法国景点依云温泉以及文化象征"香榭丽舍大道"作为主题,为人们提供一种享受,使身体得到放松和保养的生活模式,打造引领潮流的乐活社区。

规划中的唐山湾游艇社区项目,将引用欧美式Rhodes Island Nowport的建筑及景观风格。这是一种古典的建筑风格,可以适应北方严峻的气候条件,用世界闻名的港口,为富有的度假人士建造了华丽欢乐的度假圣地。

URBAN DESIGN
OF THE EAST ENTRY OF TAIHU LAKE INTERNATIONAL TOURISM ZONE, SUZHOU
苏州太湖国家旅游度假区东入口地区城市设计

Client : Management Commission of Taihu National Tourism Zone, Suzhou
Location : Suzhou Taihu National Tourism Vacation Zone
Land area : 373ha
FAR : 0.8
Building area : 1 000 000 m²
Height : 100m
Function : hotel, apartment, SOHO, office, retail, outlet and so on
International bidding : International bidding project

客　　户：苏州太湖国家旅游度假区管理委员会
位　　置：苏州市太湖旅游度假区
用地面积：373hm²
容 积 率：0.8
建筑面积：100万m²
建筑高度：100m
主要功能：酒店、公寓、SOHO、办公、商业、OUTLETS等
国际竞标：国际竞标

To build a portal node
As this project is the window of the entire tourism vacation area, the building of an important node will be the most noticeable feature of the region. A portal is not just limited to a series of "points". It is more about lines that extend to form faces. The design takes tourists and residents as the starting points and attempts to offer them an opportunity to experience the individuality of the portal area and the beauty of the city.

To extend the ecological tourist city
As an important part of the residential park of Suzhou Taihu national tourism vacation area, this region will expand the whole ecological resources of Suzhou, striking an impressive character of the whole tourist area.

打造门户节点
本项目作为整个度假区对外的第一张名片，重要节点的打造将成为该地区最显著的特征。门户不仅仅局限于一个个的"点"，更注重一条条的"线"，继而延伸至一个个的"面"，以旅游者和居住者为出发点，使他们在度假区内感受门户区的个性及城市风华。

延伸生态旅游城
该地区作为苏州太湖旅游度假区的居住园区的重要组成部分，将会拓展整个苏州生态环境资源，为整个旅游度假区树立一个强有力的个性特征。

DESIGN STRATEGY 设计策略

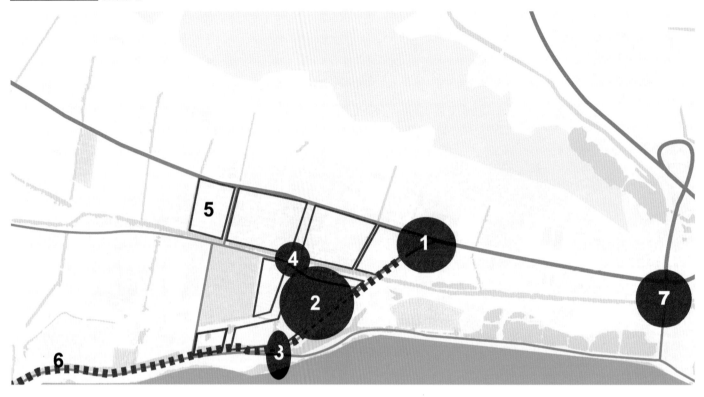

1. 创造个门户
   标志性建筑、信息中心、景观、停车和人行
2. 打造景观节点
   打造人行区域的活力和吸引力、加强视线与水系的联系
3. 交通口的处理
   人行及车行、从文化休闲商业用地到太湖边、太湖边到文化休闲商业用地
4. 在地铁站创造城市广场
   满足商业及混合使用、充分考虑各种地下功能用途
5. 研究服务型公寓地块
6. 各个旅游道路的剖面
   从景观，公交，停车，轻轨等角度出发
7. 打造联系苏州城区的门户

通过修改了部分地块的面积和道路系统，我们得到了以上7点设计策略内容，我们力求：更容易实现、更简单、更有价值

1. Greate the gateway
   Landmark building , Information center , Landscape , Crossing and parking
2. Create the destination by a signal
   Pedestrian area:Village of attraction / activities , Strong relation with view and water
3. Treatment of thecrossing road
   For pedestrian and drivers, From the touristic village to Taihu , From Taihu to the touristic village
4. Create the urban plaza at the subway station
   Mix use and commercial,Increasing density and work with complexity
5. Develop the block by service apartment
6. Qualify the section of the touristic road
   Visual axis by landscape,light,buses parking
7. Create the gateway connecting to Suzhou

This different points of the strategy are thinking according to the modification of the blocks area and road system to be:More simple, More easy to do it, More valuable

CONCEPT OF GATEWAY 门户的定义

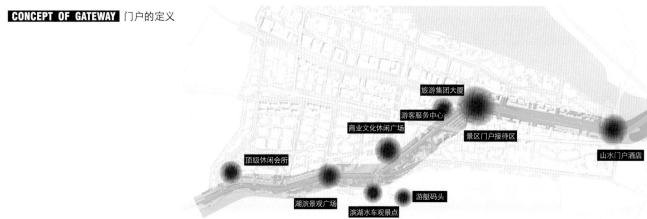

SKYLINE ANALYSIS 天际线分析

南立面
South Facade

东立面
East Facade

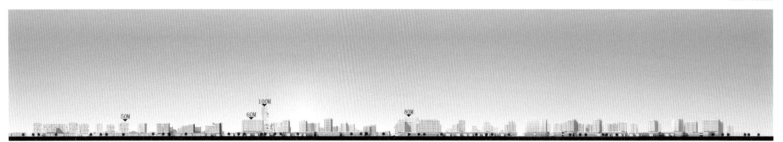

北立面
North Facade

065 · URBAN PLANNING

中心广场鸟瞰图
CENTER PLAZA BIRDVIEW

核心区鸟瞰图
CORE AREA BIRDVIEW

沿孙武路人视图
SUNWU ROAD PERSPECTIVE

入口"门户"人视图
GATEWAY ENTRANCE PERSPECTIVE

PLANNING&ARCHITECTURAL
DESIGN OF B-TEC CITY OF SHENZHEN BAY
深圳湾生态科技城B-TEC项目规划及建筑方案设计

Client: Shenzhen Investment Holding Co., Ltd.
Location: Nanshan District, Shenzhen
Total land area: 20.32 ha
Total Building area: 1 200 000 m²
International bidding: runner-up of Bid Section - 2

客　　户：深圳市投资控股有限公司
位　　置：深圳市南山区
总用地面积：20.32hm²
总建筑面积：120万m²
国 际 竞 标：二标段第二名

069 · URBAN PLANNING

BID SECTION 1 一标段

Land area: 5.8 ha

FAR: ≤6

Building area: 250 000 m²

Height: ≤100m

Function: industrial space, apartments, hotel, retail

用地面积：5.8hm²

容积率：≤6

建筑面积：25万m²

建筑高度：≤100m

主要功能：产业用房、公寓、商务酒店、商业

BID SECTION 2 二标段

Land area: 7.3 ha

FAR: ≤6

Building area: 350 000 m²

Height: ≤100m

Function: industrial space, retail

用地面积：7.3hm²

容积率：≤6

建筑面积：35万m²

建筑高度：≤100m

主要功能：产业用房、商业

DESIGN CONCEPT FOR BID SECTIONS -1& -2 一二标段设计理念

The design derives from the root of harmonious nature and injects it to the park, creating an experiential retreat and pleasant space for work, life and leisure at the same time. The architectural design stresses unitization and modulization, saving construction cost and time. The shared space enables organic combination of unit modules, thus resulting in the extending and spreading of all modules and their taking in of the best outside resources. The design also leaves the modular space free and complete. The maximization of service efficiency emphasizes inter-modular relations and creates an overlaying space between the inside and the outside, thus lifting the quality of the office space. The insertion of balconies and vegetation into the modular space serves to shield the architecture from harmful sunlight and noise.

设计萃取大自然万物和谐共生的源质，注入园区，为人们创造真实的回归体验，打造工作、生活、休闲一体化的宜人空间。

建筑设计强调单元化与模数化，节约建造成本与建造周期。利用共享空间将单元模块有机地组合，每个模块尽可能地舒展开敞，获取最佳外部资源。

模块空间自由完整，最大化使用效率，强调模块之间的对接和联系，产生相互交错的内外空间，提升办公空间品质；模块空间引入阳台与植被，提供遮阳与削弱噪声的功能。

BID SECTION 1 一标段

CONCEPT 规划设计概念

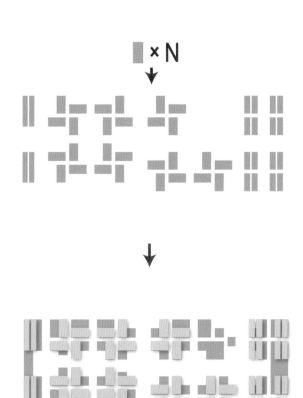

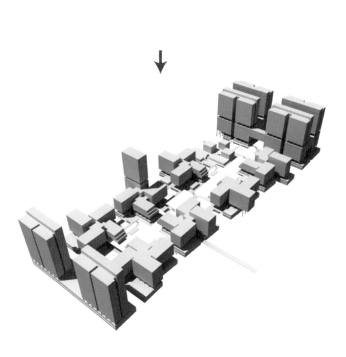

BID SECTION 2 二标段

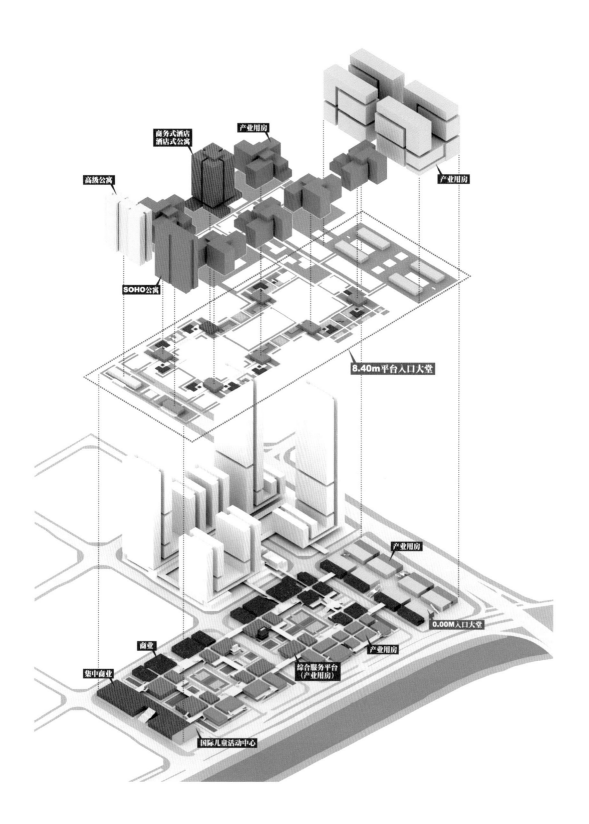

BID SECTION 3 三标段

Land area: 4.65 ha
FAR: 6.41
Building area: 298 100 m²
Height: 180 m
Function: industrial space, retail, bus station

用地面积：4.65hm²
容 积 率：6.41
建筑面积：29.81万m²
建筑高度：180m
主要功能：产业用房、配套商业、公交车站

BID SECTION 4 四标段

Land area: 2.57 ha
FAR: 11.6
Building area: 299 200 m²
Height: 250m
Function: industrial space, high end office, five star hotel and retail

用地面积：2.57hm²
容 积 率：11.6
建筑面积：29.92万m²
建筑高度：250m
主要功能：产业用房、高端办公、五星级酒店及配套商业

DESIGN CONCEPT FOR BID SECTIONS -3& -4 三四标段设计理念

The scheme sets about from general planning and considers the architectural shape and layout comprehensively from the vintage point of ecology and energy saving. It takes into account both the panoramic concept and the distinctiveness of the said sections, using prominent nodes to project their charisma. The main building follows the idea of high – quality, new shape and sustainable development. In terms of ecology and energy saving, the combination of low – input energy saving measures and high – input advanced technologies successfully reaches the goal to erect a multi – dimensional city and ecosystem demonstration park.

从总体规划着手，从生态节能的角度全面考虑建筑的形态与布局，以全标段统一设计的全局概念为基本原则进行思考，同时针对本标段的特性，设计具有特色的节点，凸显本标段的吸引力。在主体建筑设计上，呈现高品质、新形态、可持续发展的空间；生态节能方面，使用低投入节能措施为主，高投入高端技术为辅，实现建造立体多维城市，生态示范园区的目标。

BID SECTION 3 三标段

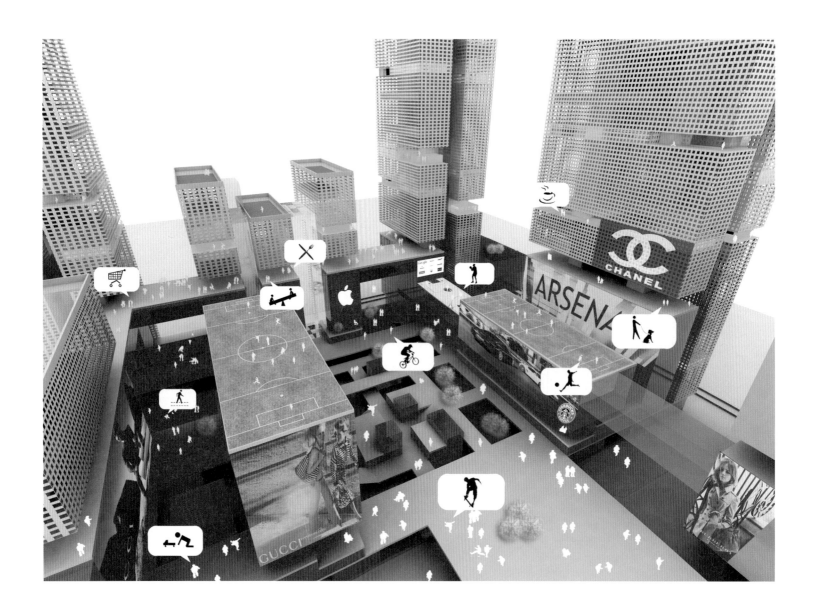

081 · URBAN PLANNING

BID SECTION 4 四标段

PLANNING&ARCHITECTURAL
DESIGN OF GREAT-WALL COMPLEX BUILDING IN SUNGANG, SHENZHEN
长城笋岗城市综合体规划及建筑概念方案设计

Client : Shenzhen the Great Wall logistics Co., LTD
Location : Sungang Centre, Luohu District, Shenzhen
Land area : 6.65ha
FAR : 5.57
Building area : 371 000 m²
Height : 150 m
Function : logistics, business, office buildings, hotels, apartments
International bidding : international bidding project

客　　户：深圳市长城物流有限公司
位　　置：深圳市罗湖笋岗中心区
用地面积：6.65hm²
容 积 率：5.57
建筑面积：37.1万m²
建筑高度：150m
主要功能：物流、商业、写字楼、酒店、公寓
国际竞标：国际竞标

The concept is of multilayer images.

The inherent logic of business lies in the gradual decrease in traffic, people and merchandises. Hydrotherapy is a cure to commercial acrophobia, and waterfall-type business resulting from the upgoing traffic and the hanging parking lot makes multilayer stacking a reality. Passages and gates being crucial, a successful business is supposed to be both a desirable labyrinth and a retractable container. Just as a vortex is a typhoon in water and a ripple a nebula in water, production and consumption are the two facets of survival. GREAT-WALL is different from GREAT-MALL in just one letter. One stone is of the capacity to make a big splash----this is the depiction of architectural morphology, our best wish, and most of all, our promise to cities.

设计概念具备多重意象。

车流、人流、物流的逐级递减是商业流的内在逻辑。水疗是治愈商业恐高症的良方，车流上行与空中停车场催生的瀑布型商业使多首层叠加得以可能。人流的路径与闸口至关重要，成功的商业理应是人乐在其中的迷宫和收放自如的容器。漩涡是水中的飓风，涟漪是水中的星云，生产—消费是生存的两个面相。"一石激起千层浪"，既是建筑形态的描摹，又是美好的祝愿，更是我们对城市的承诺。

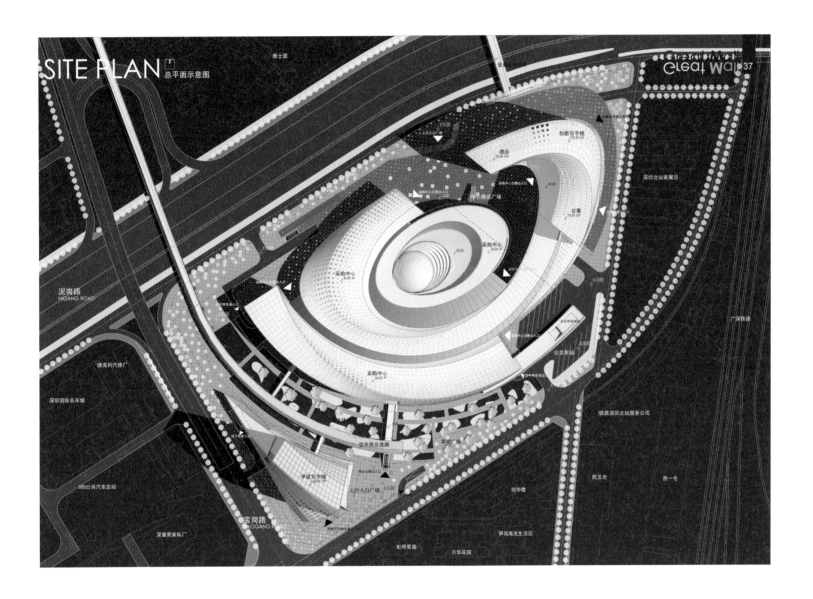

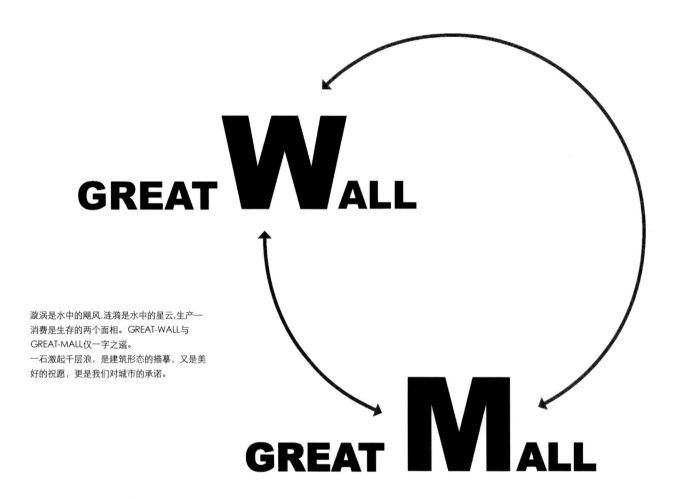

漩涡是水中的飓风,涟漪是水中的星云,生产—消费是生存的两个面相。GREAT-WALL与GREAT-MALL仅一字之遥。
一石激起千层浪,是建筑形态的描摹,又是美好的祝愿,更是我们对城市的承诺。

087 · URBAN PLANNING

PLANNING&ARCHITECTURAL
DESIGN OF THE YUHANG HIGH-TECH INDUSTRY PARK (PHASE I), ZHEJIANG
浙江省余杭高新技术产业园区科技园一期工程规划及建筑设计

Client : Yuhang Economic Development Zone, Hangzhou
Location : Yuhang District
Land area : 16.1ha
FAR : 1.6
Building area : 392 000m²
Height : 60m
Function : science & technology park
International bidding : winning project

客　　户：杭州市余杭经济开发区
位　　置：余杭区
用地面积：16.1hm²
容 积 率：1.6
建筑面积：39.2万m²
建筑高度：60m
主要功能：科技园
国际竞标：中标方案

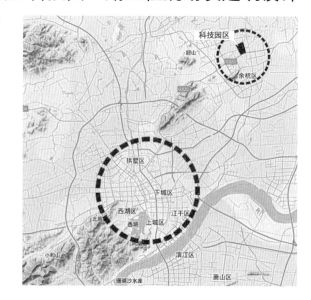

Improvement on planning

The added landscape axes interweave with the central landscape zone, changing the closed and single-row layout in the past and making more buildings capable of enjoying the beautiful scenery of the central park directly. Well-arranged distribution of buildings increases the distance between buildings and reduces the possibilities of direct eye contacts. The buildings are integrated through roofs and corridors, creating a strong sense of continuation.

Concept of creative corridor

The planning introduces the concept of creative corridor. Taking the central landscape as the main axis, it designs a wind and rain corridor running through the whole park and accordingly, linking the park area with the city space and natural hills in an organic way.

Concept of ecological axis, water axis and green axis

Taken as the themes of the central landscape axis, water and greenbelt permeate every courtyard and inner part of buildings, creating various places for communication and relaxation and realizing the genuine harmonious unity between people and nature.

规划提升

规划中增加了多条景观分轴线与中央景观带形成互相渗透的关系，改变原来较封闭和单一行列式的布局方式，使更多的建筑单体能直接欣赏到中央花园的优美景色。错落有致的建筑布局有利于增加建筑间距，减少相互对视。通过屋顶和连廊把各地块连成一个整体，建筑间实现共享，营造一种强烈的连续感觉。

创意长廊的概念

规划中引入了创意长廊的概念，以中央景观为主轴，设计一个风雨长廊体系，贯穿整个园区，使园区与城市空间和自然山体之间形成有机的联系。

生态中轴、水轴、绿轴的概念

中央景观轴中以水和绿地为主题，并渗透到各组团庭院和建筑内部，形成了多种多样的交流和休闲场所，使人与自然达到真正的和谐统一。

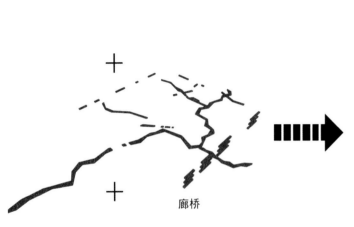
廊桥

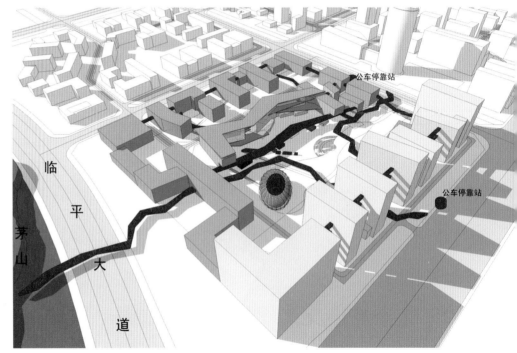
茅山　临平大道　公车停靠站　公车停靠站

透视图

透视图

PLANNING&ARCHITECTURAL
DESIGN OF THE LOT 1-C2-03 OF ZHUANTANG AREA, HANGZHOU
杭州转塘(1-C2-03地块)项目规划及建筑设计

Client : Hangzhou Rongxing Real Estate Co., Ltd
Location : Zhuantang, Hangzhou
Land area : 6.14ha
FAR : 1.81
Building area : 111 000m²
Height : 50m
Function : office, retail, apartments with hotel-styled services
International bidding : international bidding project

客　　户：杭州荣兴置业有限公司
位　　置：杭州市转塘
用地面积：6.14hm²
容 积 率：1.81
建筑面积：11.1万m²
建筑高度：50m
主要功能：办公、商业、酒店式公寓
国际竞标：国际竞标

For Chinese people, a courtyard is not just a way of residence. What's more, it is a kind of culture complex which contains in itself the relationship among humans and that between humanity and nature. It is the uniqueness of the design that traditional courtyards are incorporated into modern blocks of multiple functions, adding modern atmosphere to them.

It is convenient for small creative enterprises to rent detached houses enclosed by traditional courtyards whose independence and flexibility of usage are guaranteed. The overhead space at the bottom can provide more room for public activities and combine with business at the bottom into various commercial squares.

The flexible small commercial blocks can be easily rent as a whole or separated into single rooms. Besides, the flexible layout attracts more people, adding more fun to shopping. The theme square helps to gather crowds and serves as a label of the commercial district. Also, different themes can be used at different periods to maintain the constant attractiveness of the commercial blocks.

院落对中国人来说不仅仅是一种居住方式，更多的是一种文化情结，是在院落中寄托的人与人，人与自然之间的关系。在现代多功能复合的地块中，引入传统的院落，使之具有现代的风貌，也是这次设计中探讨的方向。

传统院落围合的独立院子，方便小型创意企业租用，保证独立性和使用的灵活性。底层架空空间能够提供更多的公共活动空间，与底层商业结合形成不同的商业广场。

灵活的小体块商业，整体租用或是划分成单间都相对方便，同时灵活的布局也有利于提升人们逛街的乐趣，吸引更多人流。主题广场有助于形成人群的聚集点，提升整个商业区的标识性，同时还可以根据不同时期更换不同主题，保持商业街的持续吸引力。

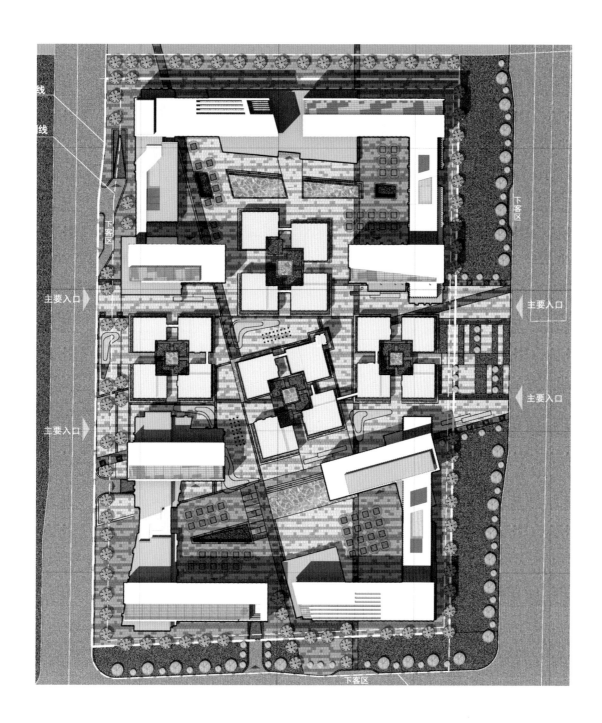

基本模块

方案基本模块由两种大小的院子各4个组成，基本模块形态完整但却缺乏相应的灵活性。

开放边缘

先打开4个大型院子的一边，使其相互对内开放，组合成更大的内部合院。形成相对独立的南区和北区。

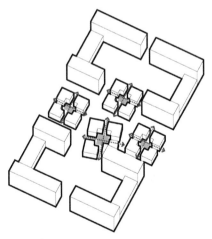

灵动内院

对小院子进行切分和旋转，在保证内院的完整性同时，也为外部提供更多交互的可能性。

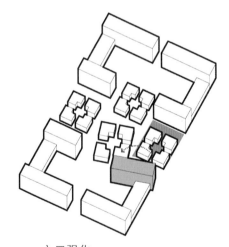

入口强化

杭富路作为城市主干道，考虑入口形象问题，将靠近杭富路的一个板楼向内侧旋转，形成扩大的主入口。

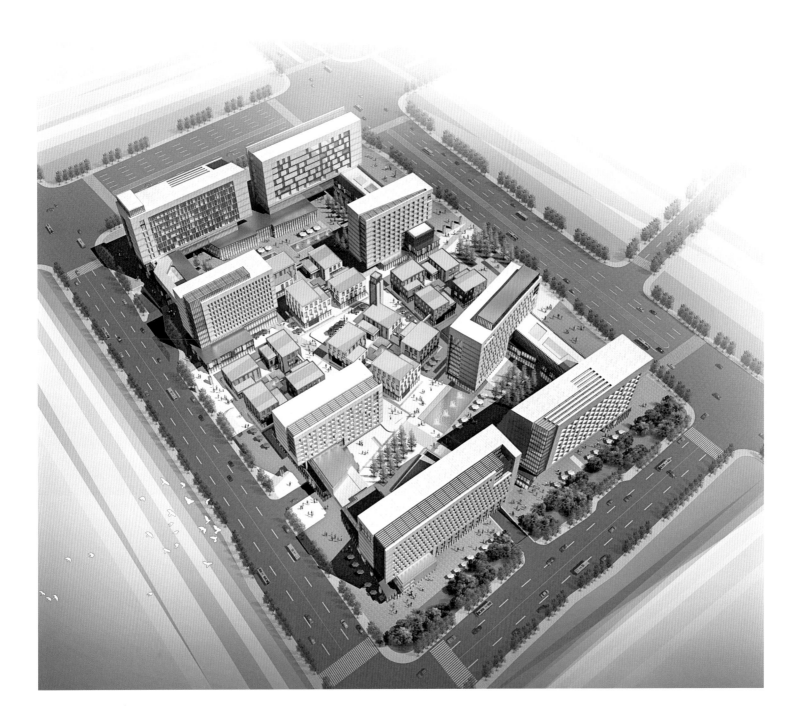

AUBE2011

AUBE2011

ARCHITECTURAL
DESIGN OF THE PARCEL A OF TANGLANG ROLLING STOCK DEPOT, SHENZHEN
深圳市塘朗车辆段A区建筑概念方案设计

Client : Shenzhen Metro Group Co., Ltd.
Location : Nanshan District, Shenzhen
Land area : 4.35ha
FAR : 6
Building area : 261 000m²
Height : 200m
Function : office, hotel, apartments, residence and retail
International bidding : 1st prize

客　　户：深圳市地铁集团有限公司
位　　置：深圳市南山区
用地面积：4.35hm²
容 积 率：6
建筑面积：26.1万m²
建筑高度：200m
主要功能：办公、酒店、公寓、住宅、商业
国际竞标：第一名

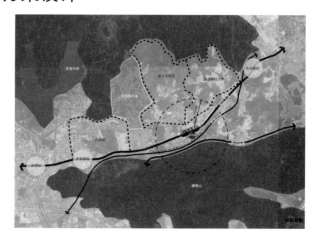

Located above the Tanglang subway, this project is an urban complex of multi-functions like office, hotel, apartments and retail.
The major design principle of the project is to optimize the surrounding urban environment and coordinate itself with the Tanglang Ecological Reserve, the college town and indemnificatory houses.
Serving as a link between the past and the future, the project is expected to integrate local resources, improve life quality of the Tanglang depot and make itself a local landmark.
Faced with complicated urban traffic, the design endeavors to reasonably arrange various traffic lines as to reduce the influence of large urban complexes on urban traffic.

项目为地铁上盖城市综合体，复合聚集办公、酒店、居住、商业等多种建筑功能。
整体设计以优化周边城市环境，协调项目与塘朗山生态保护区，大学城以及保障性住宅之间对话关系为主要设计原则。
项目应承前启后，整合片区资源，打造区域城市名片，提升塘朗车辆段工作生活品质。
面对复杂多样的城市交通，妥善安排各种交通流线，削弱大型城市综合体带来的城市交通影响。

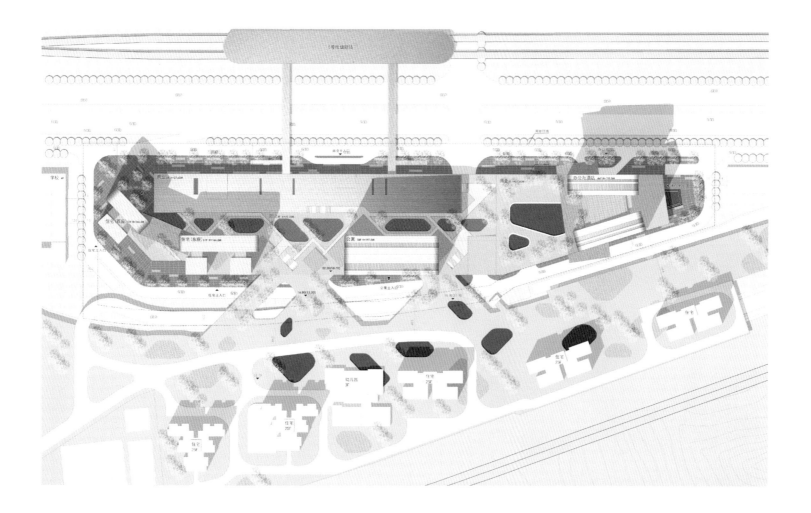

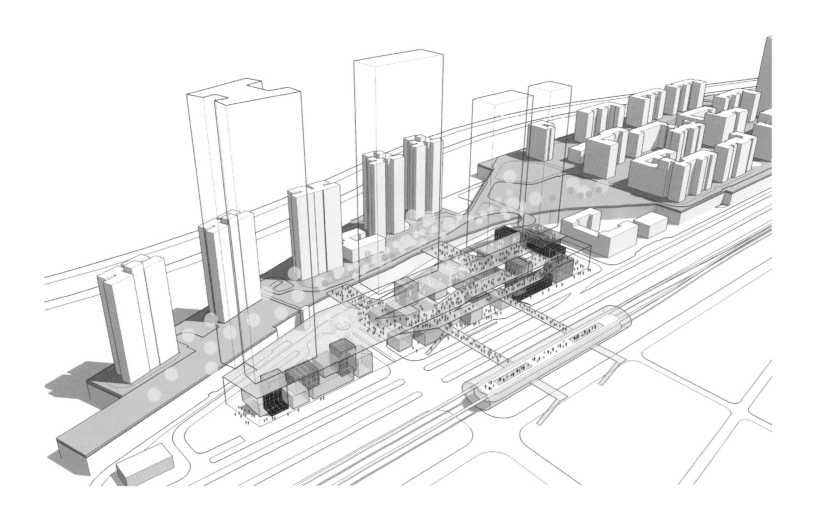

ARCHITECTURAL
DESIGN OF THE SOUTH SEA MEMORIAL HALL, FOSHAN
南海会馆概念性建筑设计

Client : Nanhai Planning Bureau, Foshan
Location : Foshan, Guangdong
Land area : 13.39ha
FAR : 0.3
Building area : 34 570m²
Height : 24m
Function : multimedia exhibition hall、salon、meeting center, etc.
Bidding project : bidding project

客　　户：佛山南海规划局
位　　置：广东佛山
用地面积：13.39hm²
容 积 率：0.3
建筑面积：3.5万m²
建筑高度：24m
主要功能：多媒体博览展馆、沙龙、会议大厅等
设计竞标：设计竞标

As the counterpart of the historical impression in reality, the South Sea Memorial Hall is labeled as a sanctuary of culture and memorial temple of spirit. In addition to the functions of a traditional memorial building, it is equipped with new compound functions. The building with diversified functions is salvaged by fishing net and placed under a big banyan, and accordingly, mountains, sea and people live in harmony.

这是一个历史印象在现实中的对应物，"文化的殿堂，精神的祠堂"是这座建筑所承载的人之理想。作为传统纪念性建筑必备的轴线，伴随着新的复合功能得以复活。纷繁多样的功能体被一张渔网打捞起来，轻放在大榕树下，于此，山、海、人相遇然后相知。

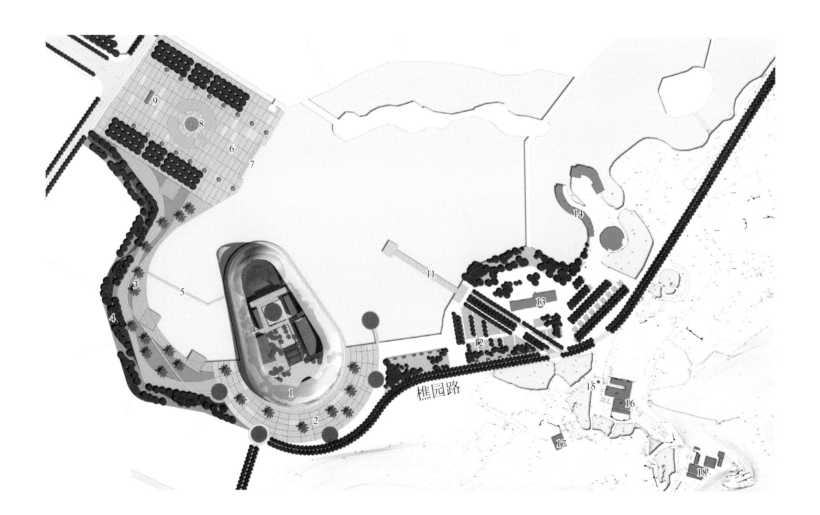

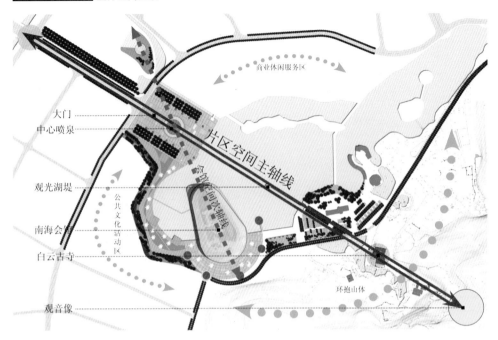

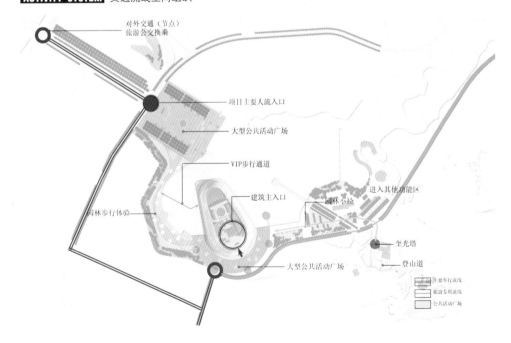

ARCHITECTURAL
DESIGN OF BAIDU INTERNATIONAL TOWER
百度国际大厦建筑设计

Client : Baidu International Science and Technology (Shenzhen)Co., Ltd
Location : Nanshan District, Shenzhen
Land area : 1.41ha
FAR : 12
Building area : 171 200 m^2
Height : 179.4m
Function : retail, office
International bidding : international bidding project

客　　户：百度国际科技（深圳）有限公司
位　　置：深圳市南山区
用地面积：1.41hm^2
容 积 率：12
建筑面积：17.12万m^2
建筑高度：179.4m
主要功能：商业、办公
国际竞标：国际竞标

The design draws inspiration from the opposition and complementarity of the size, space and texture of the two buildings. A building order is formed in conflict and fusion.

Technical and sensitive game rules bring a variety of expressions about social vitality: welcome, exchange, dialogue, surprise, exploration and interaction. There is no absolute order among these expressions which all endow things with idealized forms. Only some initial movements and lines are stated, annotated, set and constructed in the first place, and they are sometimes integrated, sometimes scattered, sometimes focused, sometimes open to the sky……

设计构思来源于两个建筑物的体量、空间和肌理的对立性和互补性上，在冲突与融合的过程中构建了一种建筑次序。

技术性的、敏感的游戏规则开启了多种社会活力的表达：迎接，交换，对话，惊喜，探索和相互联系。所有表达并不存在一个绝对的次序，并都能给予事物理想化的形式。只有最初的一些运动、线条，首先被申明，被注释，被安置和建造，时而整合，时而分散，时而集中，时而向天际开阔……

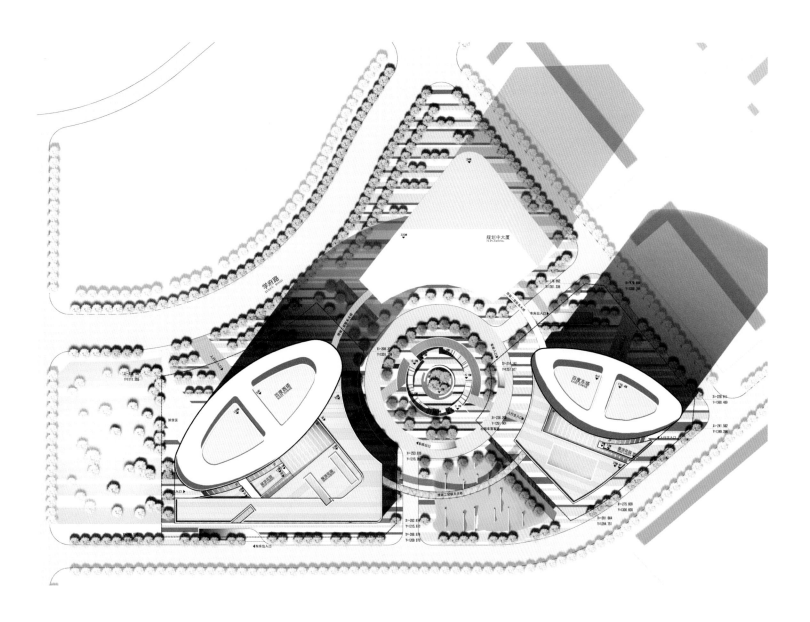

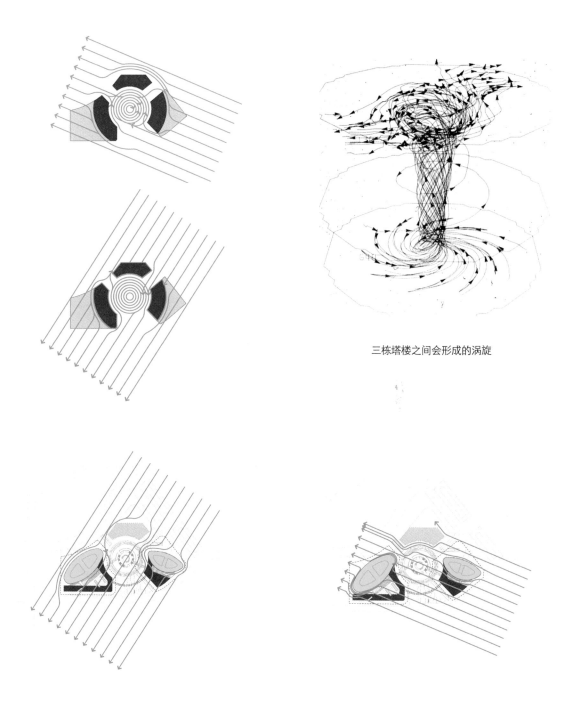

三栋塔楼之间会形成的涡旋

除了良好的展示型和开放视野，塔楼的形状和布局，也是考虑到优化三栋塔楼之间的风环境，使气流趋于平缓，避免广场上形成涡流干扰使用的安全和舒适。

FUNCTION LAYOUT 功能分布

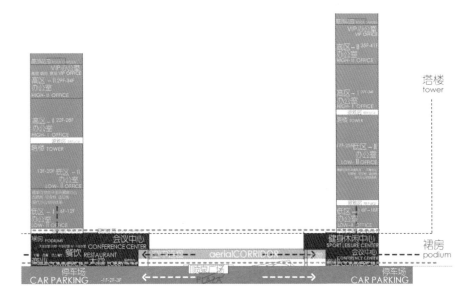

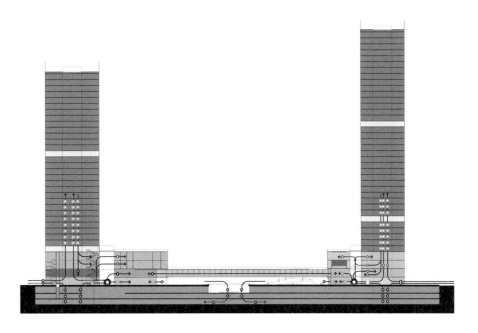

ARCHITECTURAL
DESIGN OF THE UPGRADING & RENEWAL PROJECT OF THE INDUSTRIAL PARK IN THE NORTH OF OCT, SHENZHEN
深圳市侨城北工业区升级改造工程建筑设计

Client : Wan Thai Real Estate(Shenzhen) Co., Ltd
Location : Nanshan District, Shenzhen
Land area : 4.88ha
FAR : 4.6
Building area : 307 300m²
Height : 150m
Function : Industrial office, dormitory
International bidding : winning project

客　　户：运泰建业置业（深圳）有限公司
位　　置：深圳市南山区
用地面积：4.88hm²
容 积 率：4.6
建筑面积：30.73万m²
建筑高度：150m
主要功能：产业办公、配套宿舍
国际竞标：中标方案

Green plants are used to create benefits. With the building of green space taking a dominant position, the design is to induce the upgrading of green plants and creates a three-dimensional urban ecological community.

A pleasant environment of low density forms under the condition of high density and the flexible boundary and human scale of buildings soften their boundary with the environment. Breathing is the basis of conception. The green island, just like lungs, provides for emerging industries a vehicle of conception and incubation, and also represents our expectation of a glorious future.

The layout of open neighborhoods puts emphasis on paths and feelings of people, forming a pleasant maze and flexible vessel.

点绿成金，创造效益，绿色空间营造成为设计的主导，设计要激发一次绿色的升级，创造一个城市生态立体社区。
高密度条件下塑造低密度的宜人环境，建筑的柔性边界和人性尺度软化了和环境间的界限。呼吸是孕育的基础，绿岛如肺叶，为新兴产业提供了孕育和孵化的载体，也是对其美好未来的期待。
开放街区化的场地布局，重视人流的路径与感受，是人乐在其中的迷宫和收放自如的容器。

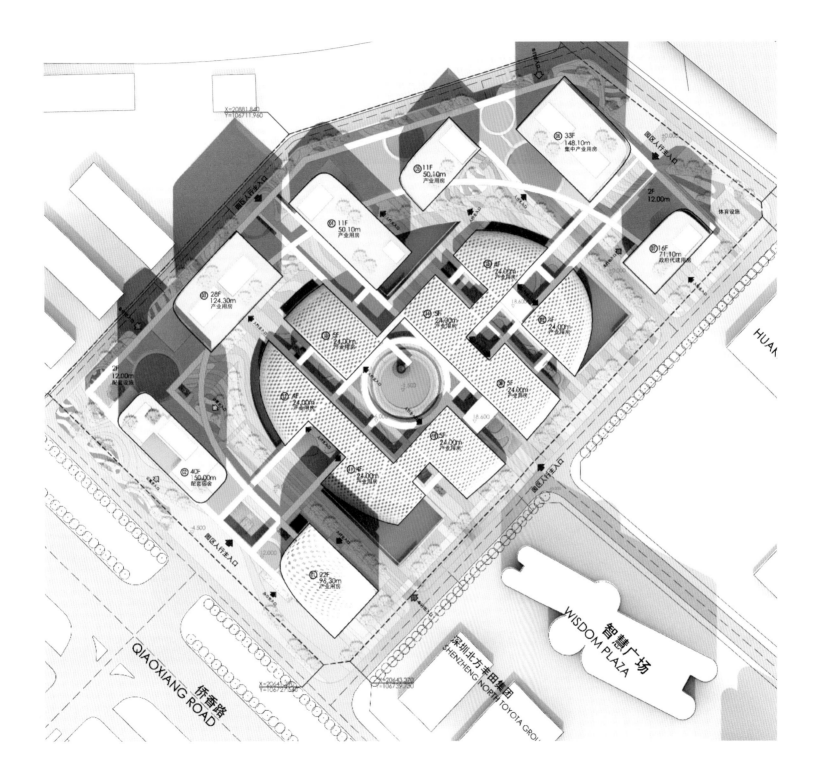

IDEA DIAGRAM 概念生成

1. 场地现状略有高差

2. 利用场地高差，设计园区主入口。

3. 形体生成

6. 添加步行连廊系统

5. 抬升高层裙房，留出更多开放空间

4. 引入十字步行轴

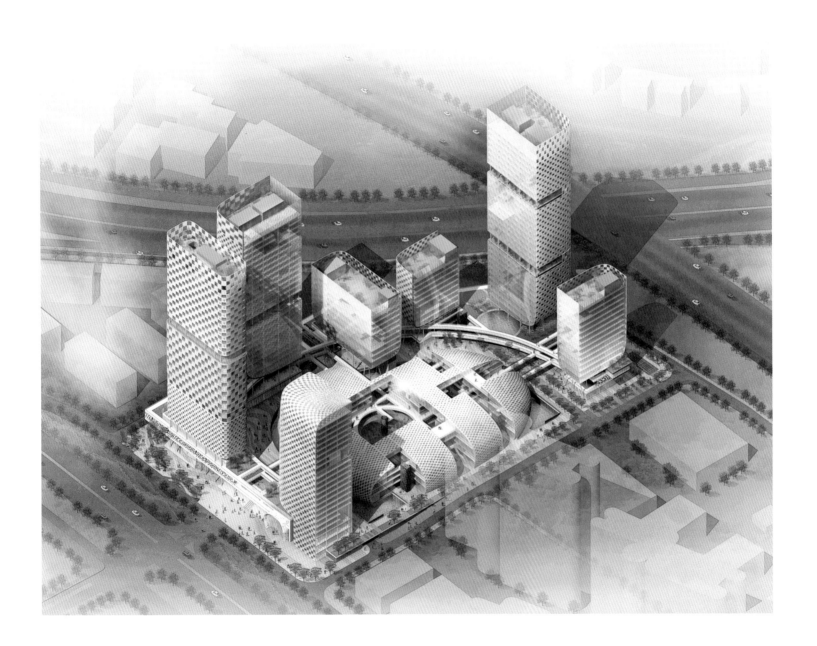

ARCHITECTURAL
DESIGN OF THE HORIZON CLUB OF EXPANDER · THE GARDEN OF BENEVOLENCE & WISDOM, SHENZHEN

城建观澜仁山智水花园天际会所建筑设计

Client : Shenzhen Urban Expander Group

Location : Guanlan, Shenzhen

Land area : 4 778m²

FAR : 0.44

Building area : 1 800m²

Height : 26m

Function : a swimming pool club for the Garden of Benevolence & Wisdom

Under construction

客　　户：深圳市城市建设开发（集团）公司

位　　置：深圳观澜

用地面积：4 778m²

容 积 率：0.44

建筑面积：1 800m²

建筑高度：26m

主要功能：小区配套游泳池会所

在建项目

Salar de Uyunia Whisper with Eternity

The designer draws inspiration from Salar de Uyuni in Bolivia, which is the largest salt marsh in the world. There are endless waters blurring the boundary between sky and earth, which quite matches this project's theme of water.

The design reconstructs the eternity embodied in the special scenery of Salar de Uyuni, symbolizing people's understanding of life inspired by the project. The word whisper indicates people's aspiration for serenity and self-reflection in a bustling society. The texture of water is quite exceptional and its changes among ice, liquid and steam show great strength. The rising icebergs are the primitive form of water, which flow s far into horizon.

天空之镜—与永恒的私语

构思源于玻利维亚境内的天空之镜，是世界最大的盐沼，如镜面一般，无尽的水面使得天空与人间的界线消失，契合本案的水主题。

提取天空之镜的特殊场景凝固的永恒感，象征本项目给予人们对生命的领悟。私语一词则道出了在烦嚣的当代社会中，人们向往的宁静与自省。水的质感一反常态，在冰、水、汽之间的强柔变化，衍生出坚定的力度。隆起的冰山宛如水的原初状态，消融后不断流淌，直至天际。

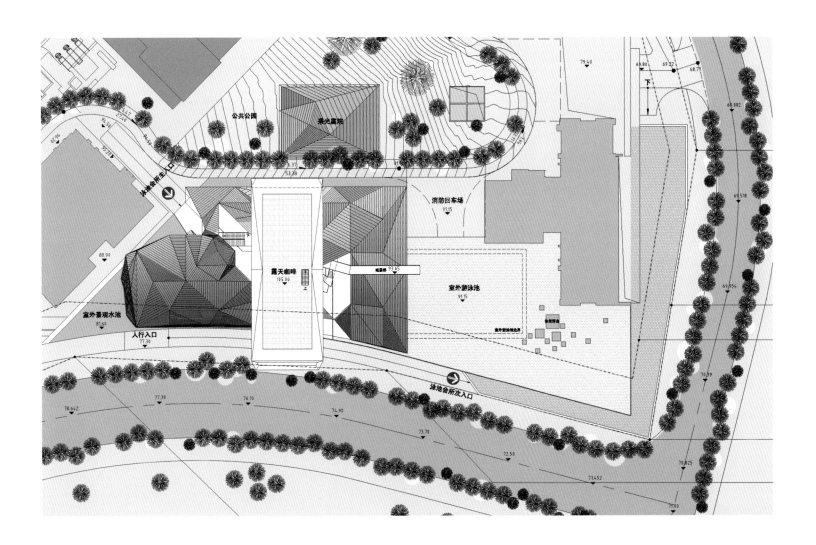

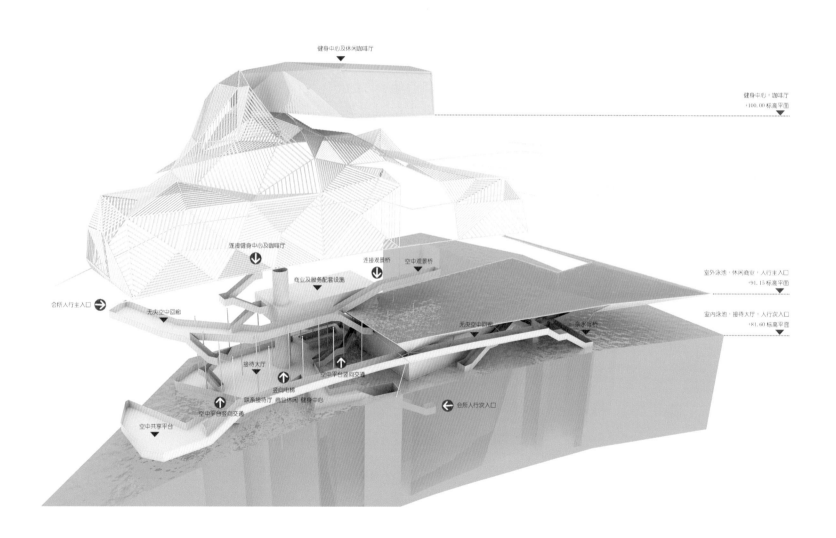

125 • ARCHITECTURE

ARCHITECTURAL
DESIGN OF THE CRSID TOWER
中铁南方总部大厦建筑设计

Client : China Railway South Investment & Development Co.,Ltd

Location : The Shenzhen Bay CBD of Houhai Central District, Nanshan, Shenzhen

Land area : 5 253m²

FAR : 8.20

Building area : 58 100m²

Height : 99.75m

Function : office, restaurants , retail

International bidding : winning project

Under construction

客　　户：中铁南方投资发展有限公司

位　　置：深圳市南山区后海中心区深圳湾金融商务区

用地面积：5 253m²

容 积 率：8.20

建筑面积：5.81万m²

建筑高度：99.75m

主要功能：办公，餐饮，商务

国际竞标：中标方案

在建项目

The design endeavors to create a steady, solid and vigorous architecture posture through reasonable plane proportion and succinct architecture language so as to manifest the characteristics of China Railway.

As for the image, podium building elevation and grey upright and energetic I-bars on north and south sides embody the connotation of the project and indicate the specialty of the building.

设计力求通过合理的平面比例以及简练清晰的建筑语言，创造出稳健扎实而又蓬勃向上的建筑姿态，诠释中国中铁的企业特质。

造型上，裙楼立面以及塔楼南北两侧采用端正、刚健的深灰色工字钢线条表达建筑力学的传递，体现工程内涵，展现建筑的特殊性。

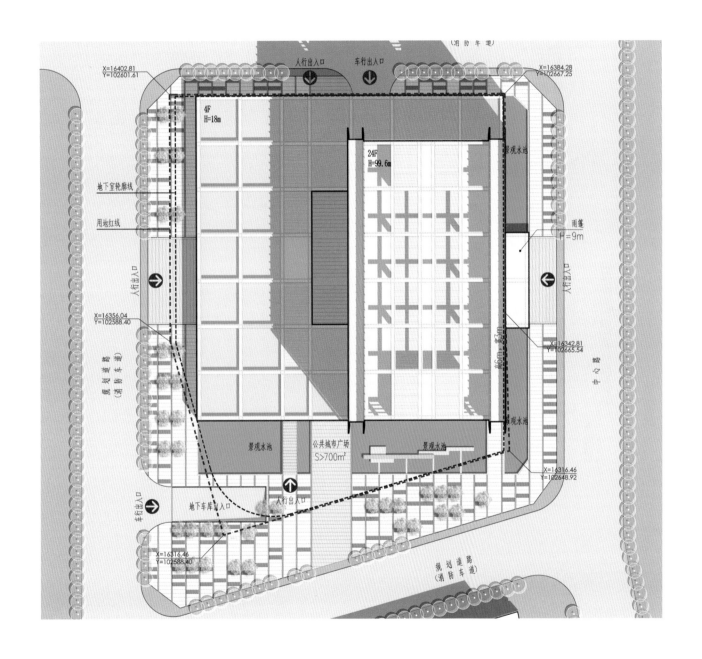

IN-CIRCULATION ANALYSIS 内部流线分析

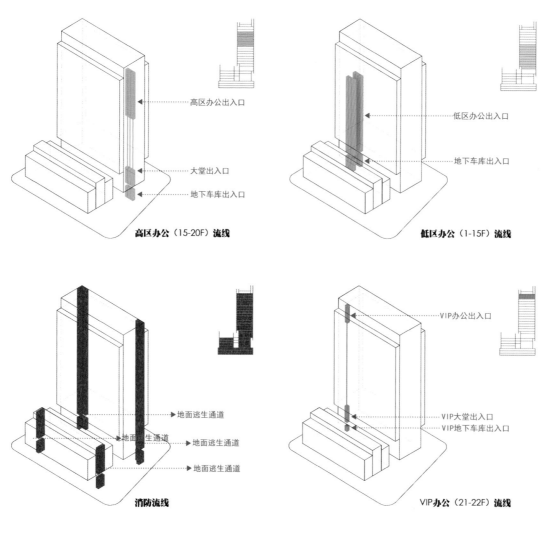
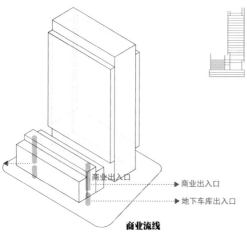

ARCHITECTURAL
DESIGN OF THE DIANCHI LAKESIDE RESIDENCE, KUNMING
昆明滇池湖岸花园项目建筑设计

Client : Yunnan Kunchi Real Estate Co. Ltd
Location : Dianchi National Tourist Resort, Kunming
Land area : 50.05ha
FAR : 0.3
Building area : 150 700m²
Height : 24m
Function : leisure and retail, outlets, headquarters offices, mountain cinema, hotel, recreation
under construction

客　　户：云南堃池房地产有限公司
位　　置：昆明市滇池国家旅游度假区
用地面积：50.05hm²
容 积 率：0.3
建筑面积：15.07万m²
建筑高度：24m
主要功能：休闲商业、奥特莱斯商业、总部办公、山体影院、酒店、娱乐文化等
在建项目

How far is Kunming, China from Paris, France?
May the former colonial history represented by Dian -Yue(from Yunnan to Vietnam) railroad be accepted esthetically and entertainingly in the post-colonial culture?
Can the Sino-French cultural exchange assume a different appearance in Kunming, Yunnan?
Is the cultural imagination and mutual possession between China and France possible?
With its symbol-wrapped multifunctions such as office, shopping, amusement, restaurants, wedding, church and sports, this project is an attempt at the consumption-centric building design.

中国昆明与法国巴黎的距离有多远？
滇越铁路书写的前殖民史在当代能否以娱乐姿态重新进入审美视野？
中法文化交流在云南昆明可否自有另外一番面目？
能否实现中国对法国的文化想象及相互占有？
设计以符号碎片包裹起办公、购物、游乐、餐饮、婚庆、教堂、运动等诸多功能，是对消费文化主导建筑设计的一次尝试。

PLANNING, ARCHITECTURAL & LANDSCAPE DESIGN OF THE PLOTS 1/2/3 OF THE COMMUNICATION INDUSTRY PARK OF QINGDAO

国家（青岛）通信产业园1、2、3号地块规划、建筑、景观方案设计

Client : Qindao High-Tec Industry Development Co.,Ltd.
Location : Laoshan District, Qingdao
Land area : 4.36ha
FAR : 4.04
Building area : 245 000m²
Height : 80-100m
Function : office, R&D, retail
International bidding : 2nd prize

客　　户：青岛高科产业发展有限公司
位　　置：青岛市崂山区
用地面积：4.36hm²
容 积 率：4.04
建筑面积：24.5万m²
建筑高度：80-100m
主要功能：办公、研发、商业
国际竞标：第二名

Within the land, main buildings are laid out in the pattern of three rows of short board along the street. A relatively long distance is left between towers so as to allow light and natural wind to enter and to form diversified internal spaces. Along the coastal avenue, buildings are beveled with the road, forming the best demonstration of the city image in the industrial garden. All buildings are constructed in minimalist style, and the facade in the modular style implicates the precision of industrial design. Following the principle of unifying buildings and landscapes, buildings rise from the ground, transform, and twist to show a "gradual rise" architectural state. Thus, the boundary between buildings and landscape is obscured and a magnificent and independent building group is created.

地块内主体建筑采用沿街三排短板布局方式，塔楼之间设置较大的楼间距，导入光线与自然风，同时形成丰富多变的内部空间。在面向滨海大道一侧,建筑与道路呈一定斜角布置，形成园区最佳城市形象昭示面。建筑本身强调简约主义风格，表达对简洁清晰的追求，模数化风格的立面意向3蕴涵了工业设计精密的思想，建筑形体遵循建筑景观一体化设计原则，裹挟着挺拔清晰像素化肌理，从地面升起、变形、回转，呈现"节节高"的建筑姿态。建筑和景观的界限被消抹，形成气势磅礴、卓然独立的建筑群体。

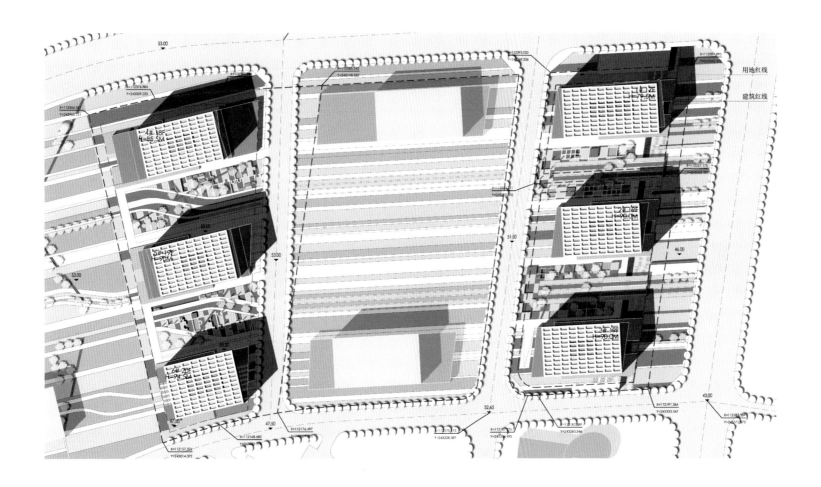

ARCHITECTURAL
DESIGN OF ALIBABA'S BRANCH TOWER IN SHENZHEN
阿里巴巴深圳大厦建筑概念方案设计

Client : Alibaba(China) Co.,Ltd.
Location : Nanshan District, Shenzhen
Land area : 1.38ha
FAR : 5.90
Building area : 108 200m²
Height : 100m
Function : retail, office
International bidding : International bidding project

客　　户：阿里巴巴集团
位　　置：深圳市南山区
用地面积：1.38hm²
容 积 率：5.90
建筑面积：10.82万m²
建筑高度：100m
主要功能：商业，办公
国际竞标：国际竞标

Considering the uniqueness of Alibaba, Alibaba's expectation for the project, conditions of the site and keypoints of the urban planning, the building is designed to be enclosed by rigid boundaries and arbitrary curves. For the more, the design pays attention to the needs of both the city and the company, and decides on a coastal architecture form with open, semi-open and private spaces.

设计从阿里巴巴公司特质，阿里巴巴集团对本项目的愿景，以及本建筑所处基地状况和城市规划要点着手研究，生成刚性边界与自由曲线围合而成的建筑形态，同时兼顾城市与企业的需求，营造出全开放、半开放与私密空间并存的迎海建筑形式。

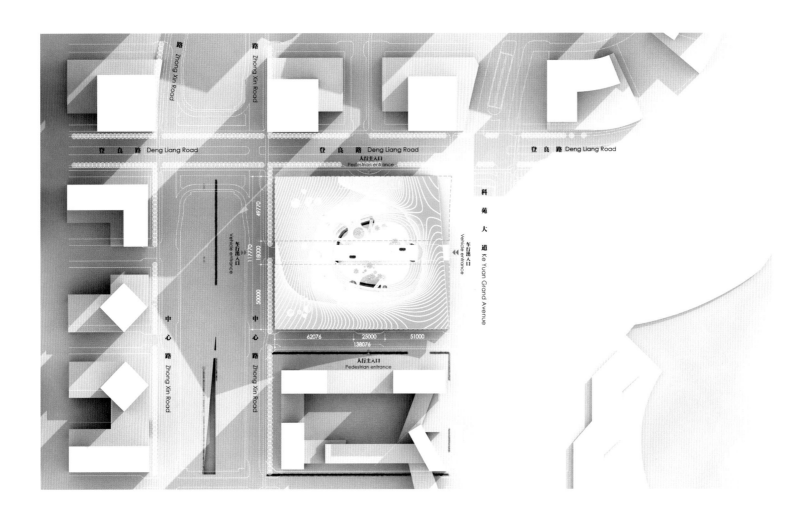

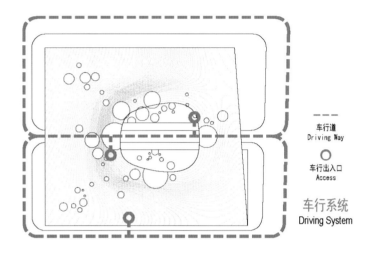

车行系统
Driving System

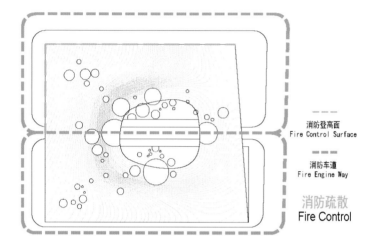

消防疏散
Fire Control

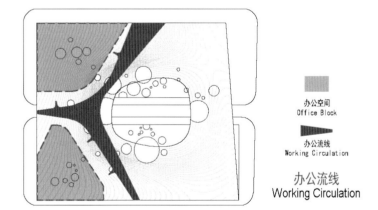

办公流线
Working Circulation

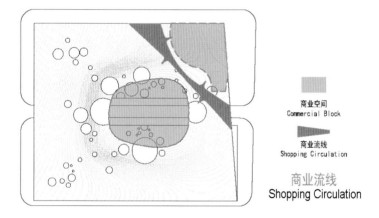

商业流线
Shopping Circulation

城市界面

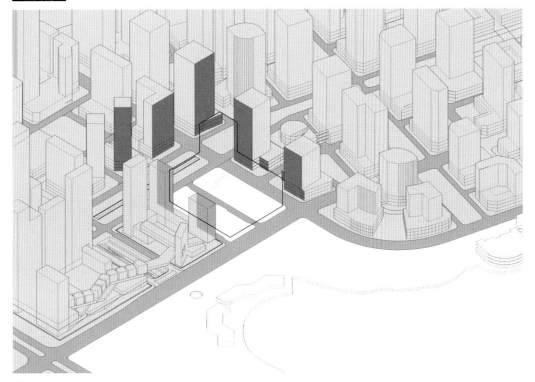

景观分析

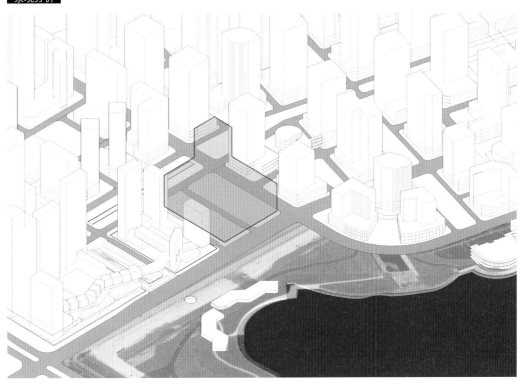

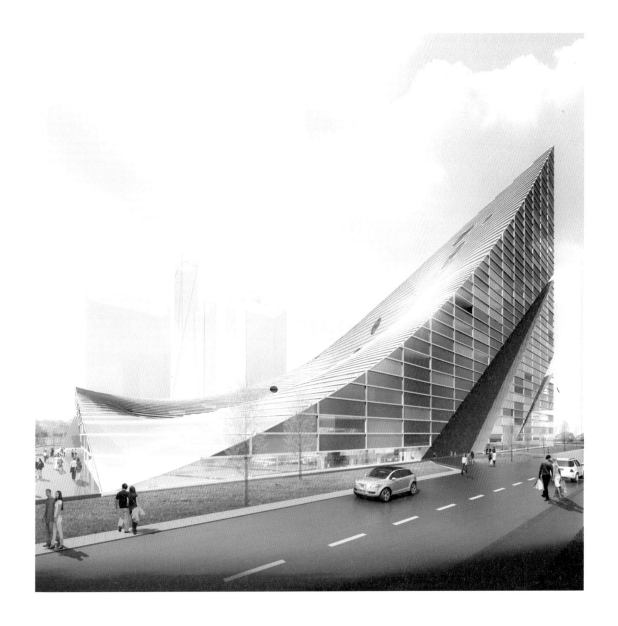

ARCHITECTURAL DESIGN OF THE RENEWAL PROJECT OF THE 34th BLOCK OF BAO'AN DISTRICT, SHENZHEN

深圳市宝安区新安宝城34区改造项目建筑设计

Client : Shenzhen Warmsun Real Estate Development Co.,Ltd.
Location : Bao'an District, Shenzhen
Land area : 1.24ha
FAR : 6.0
Building area : 74 000m²
Height : 100m
Function : retail, apartments
International bidding : winning project

客　　户：深圳市华盛房地产开发有限公司
位　　置：深圳市宝安区
用地面积：1.24hm²
容 积 率：6.0
建筑面积：7.4万m²
建筑高度：100m
主要功能：商住
国际竞标：中标方案

The concept of the general planning is "A golden commercial corridor · A full-functional life complex".
A peculiar planning concept comes into being by rendering the spatial texture of the entire Bao'an District into a modern modeling language. With a view to the whole business atmosphere along Jian'an Road, local culture and future development of the district, a 3-dimensional solution is formulated to provide a suitable residential area. A living atmosphere is cultivated in the neighborhood so as to develop a healthy and sustainable residential space.

总体规划立意为"黄金商业走廊·全方位生活综合体"。
从宝安区整体城市空间肌理出发,转译为现代模式化语言,形成独特的规划理念。结合沿建安路整体商业氛围、本地文化和地区未来发展,提供适合居住生活的全方位立体化解决方案。营造街坊邻里生活氛围,提供健康可持续发展的住区空间。

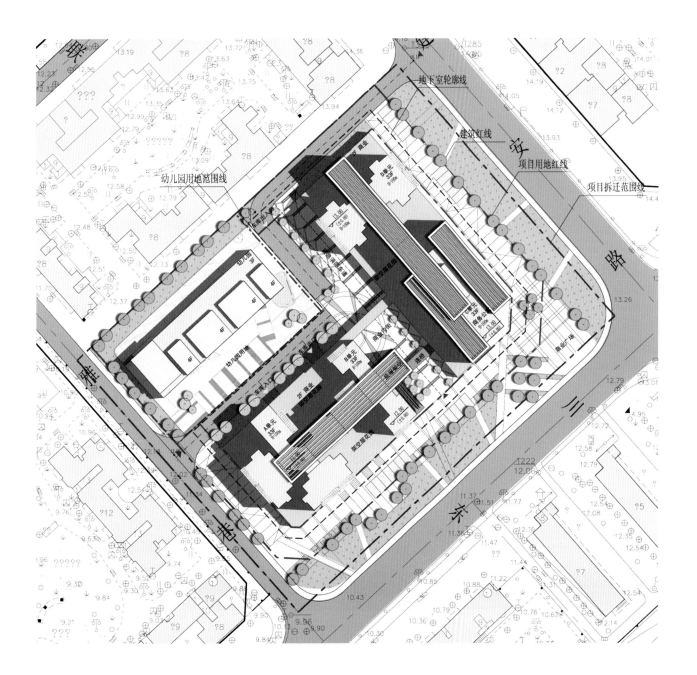

ARCHITECTURAL
DESIGN OF THE MULTIPLE-USE BUILDING IN THE WEST CAMPUS OF SHENZHEN POLYTECHNIC

深圳职业技术学院西校区综合楼工程建筑设计

Client : Shenzhen Polytechnic
Location : Nanshan District, Shenzhen
Land area : 0.52ha
FAR : 4.7
Building area : 36 000m²
Height : 60m
Function : international conference center, apartment
International bidding : 2nd prize of bidding

客　　户：深圳职业技术学院
位　　置：深圳市南山区
用地面积：0.52hm²
容 积 率：4.7
建筑面积：3.6万m²
建筑高度：60m
主要功能：国际会议中心、公寓
设计竞标：第二名

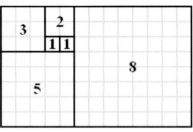

The Concept of Green Cubic

An air corridor is connected with the lush Guanlong Hill in the north, forming a green corridor which leads the mountainous greenbelts into the central yard, the open floor, hanging gardens and roof gardens, giving a full demonstration of the design philosophy of "green campus" and "ecological campus".

Fibonacci Sequence in the Design

In mathematics, the Fibonacci numbers are the numbers in the following integer sequence: （1、1、2、3、5、8、13、21、……）

By definition, the first two Fibonacci numbers are 0 and 1, and each subsequent number is the sum of the previous two（1+1=2，1+2=3，2+3=5，3+5=8……）

The ratio of consecutive Fibonacci numbers converges and the limit approaches the golden ratio 0.6180339887……

The architectural design incorporates the Fibonacci sequence and forms a unique architectural style characteristic of the golden ratio.

绿立方的概念

一条空中走廊与北面绿意盎然的官龙山相连接，形成一条绿色走廊，将山体绿化通过绿色走廊引入到中心庭院，并渗透到架空层、空中花园和屋顶花园，充分展现"绿色校园"、"生态校园"的设计思想。

在造型设计中引入斐波那契数列理论：

斐波那契数列指的是：（1、1、2、3、5、8、13、21、……）

这个数列从第三项开始，每一项都等于前两项之和（1+1=2，1+2=3，2+3=5，3+5=8……）。

随着数列项数的增加，前一项与后一项之比越来越接近黄金分割的数值0.6180339887……

建筑设计将斐波那契数列的原理融入到整个立面造型中，形成具有黄金分割比例的独特建筑风格。

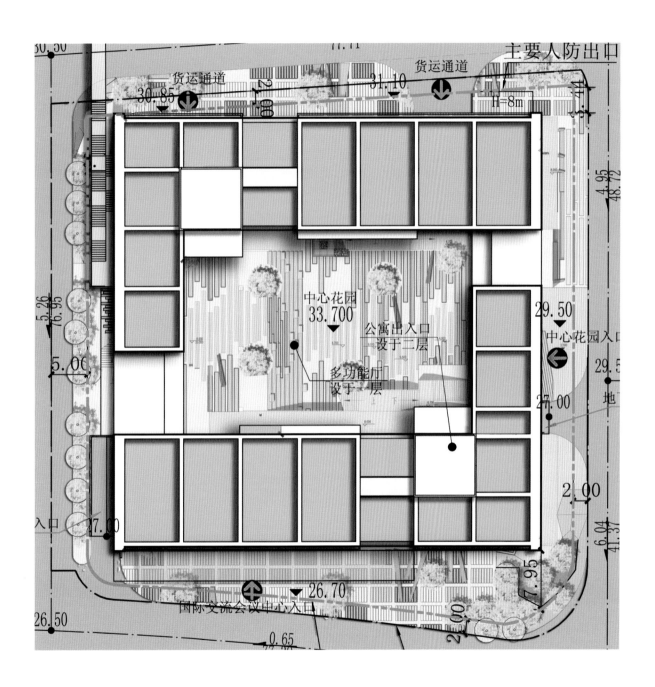

建筑体量生成

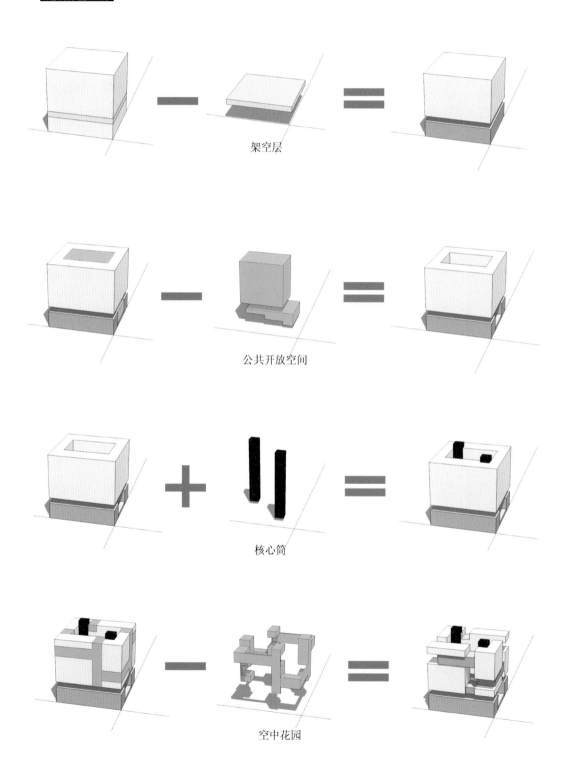

绿立方的概念

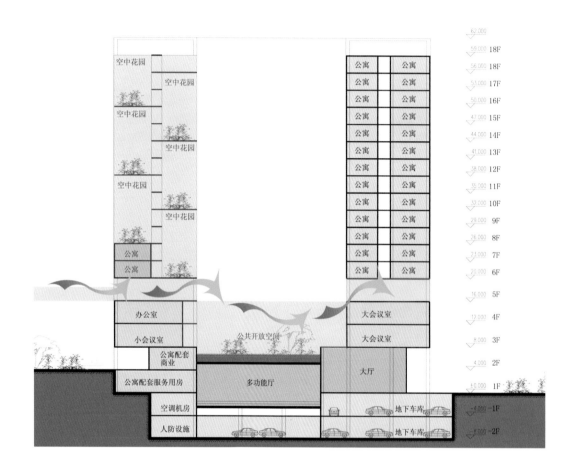

通过一条空中走廊与北面绿意盎然的官龙山相连接，形成一条绿色走廊，也为与北校区通过盘山步道的连接提供了一个可能性，将山体的绿化通过绿色走廊引入到中心庭院，并渗透到架空层、空中花园和屋顶花园，充分展现"绿色校园"、"生态校园"的设计思想，达到人与自然、建筑与自然和谐统一。

RUBE3OII

RUBE2011

景 观
LANDSCAPE

LANDSCAPE
DESIGN OF THE VICTORY GATE STATION SQUARE OF WUXI METRO LINE 1
无锡地铁1号线胜利门站广场景观方案征集

Client: Railway Traffic Planning & Construction Headquater of Wuxi Municipality
Location: Wuxi
Land area: 2ha
Landscape area: 20 000 m²
Function: landscape of traffic stop squares
Bidding project: bidding project

客　　户：无锡市轨道交通规划建设领导小组（指挥部办公室）
位　　置：无锡市
用地面积：2hm²
景观面积：2万m²
主要功能：交通站点广场景观
设计竞标：设计竞标

The design aims to create a shopping park with the theme of ribbons, flowers and hills.
The landscape design is to extend Xi and Hui Hills, bringing nature into urban life.
The main function of the landscape is to create a slow park-like shopping environment.
The hill park is constructed in the center of a modern city, which distinguishes it from the ancient and simple commercial district around Nanchan Temple.
As against the high-speed shopping environment in the Sanyang Square, a slow shopping environment with ecology, recreation, performing arts and creativity is provided, together with a multi-level space for pedestrians.
Design elements: plum flowers, link, gate
(1) Plum flowers are the city flower of Wuxi. An artificial hill is constructed in the form of plum flower, and plum flowers are planted on it to create a scene of a hill with blooming flowers in the downtown.
(2) Subways are to people in modern times what rivers are to people living in river towns. They link the city, people and life.
(3) The Victory Gate is the north gate of the ancient Wuxi city. The imagery of the gate is highlighted to create a modern gate of opposite scenery and enframed scenery.

设计旨在打造以彩带花丘为主题的情景式购物公园。
景观设计概念拟延续惠锡二山自然山脉，让自然在城市中生长。
景观功能着力创造公园式慢速购物。
打造现代城市中心的山体公园，以区别南禅寺商业古朴的水乡空间。
提供集生态、娱乐、演艺、创意于一体的慢速购物环境，多层次空间集散人流，以区别三阳广场快速的购物环境。
设计元素：梅花、纽带、门
（1）梅花为无锡市花，提取梅花形态堆筑人工山体，并遍植梅花，打造城中心山花烂漫的景观效果。
（2）地铁之于现代人，就如河道之于水乡人，如同纽带般将城市、人和生活连接起来。
（3）胜利门乃无锡古城老北门，突出门的意境，打造意断神连的对景和框景的现代之门。

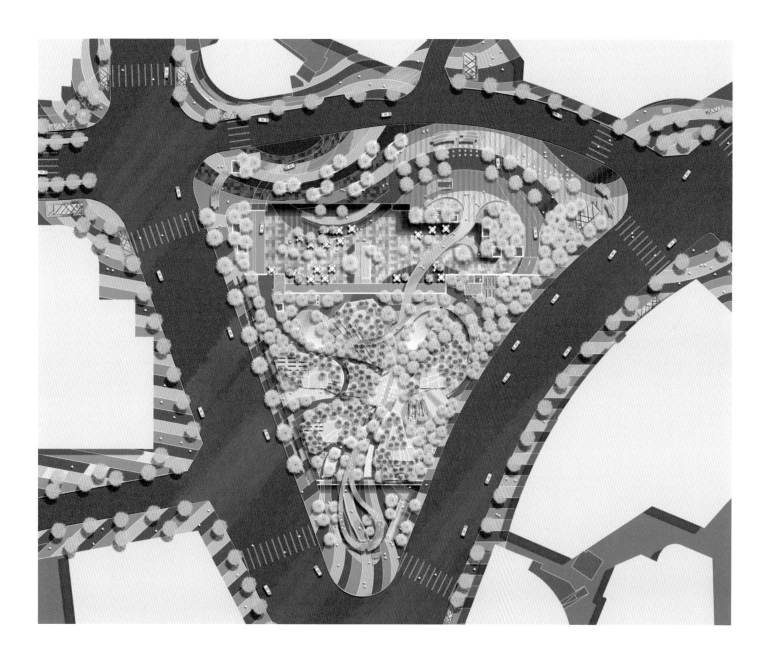

CONCEPT 设计概念

胜利之门打造城市界面，唤起历史记忆...
山水之门引入惠山山脉，打造绿色核心...
彩色飘带延续城市脉络，串连周边空间...
梅花山岭打造山花烂漫，制造视觉焦点...

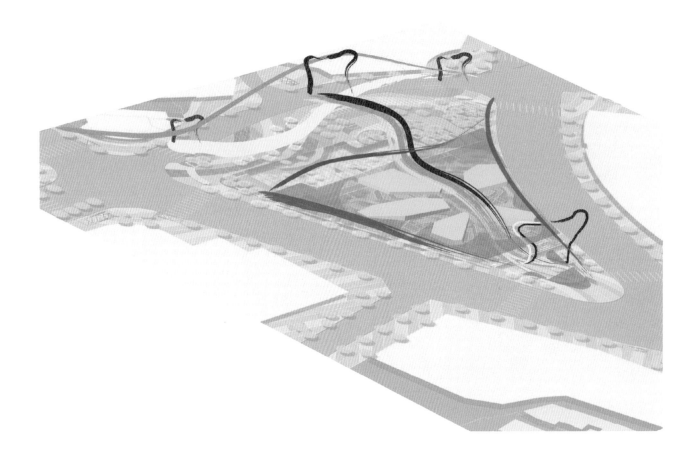

ELEMENT ANALSIS OF CONCEPTION 概念元素分析

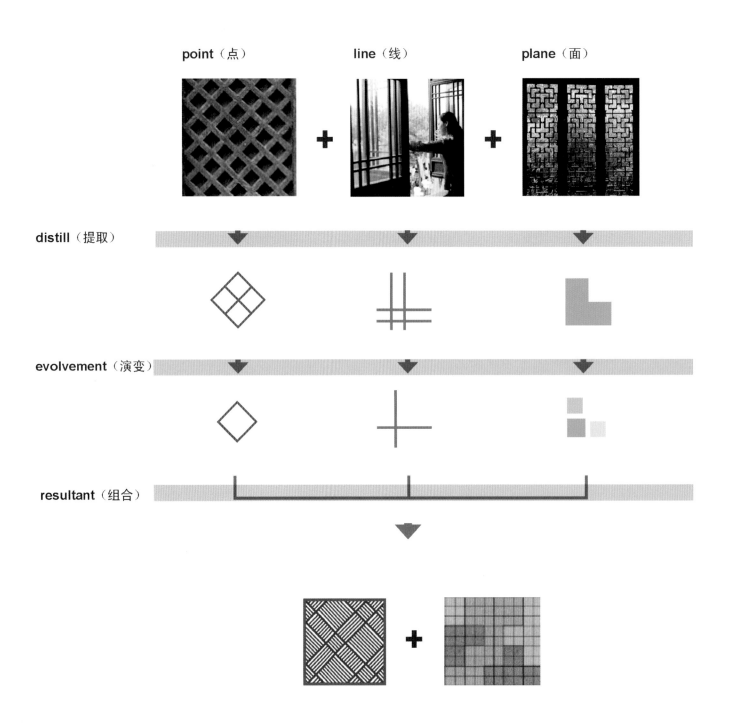

沿道路立面挡墙装饰设计

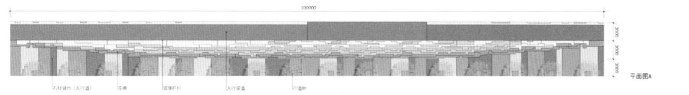

平面图A

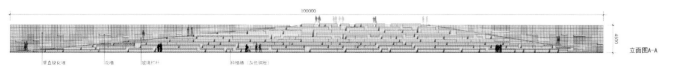

立面图A-A

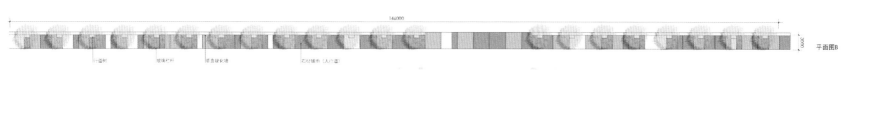

平面图B

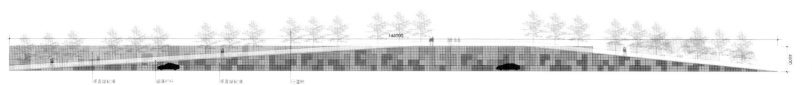

立面图B-B

"飘带"构架设计

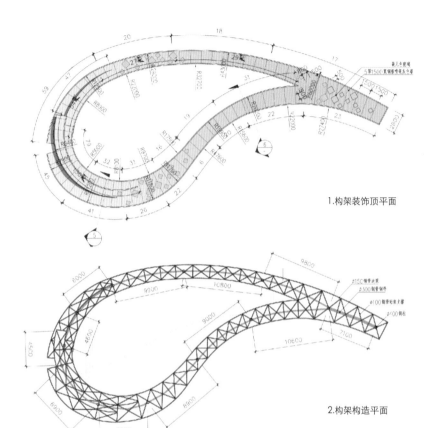

1. 构架装饰顶平面

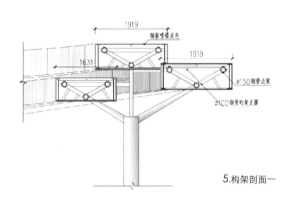

5. 构架剖面一

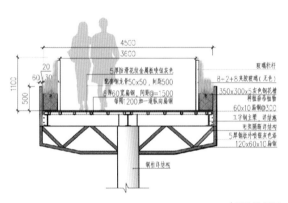

2. 构架构造平面

4. 桥构造剖面

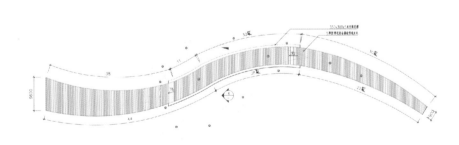

3. 桥平面

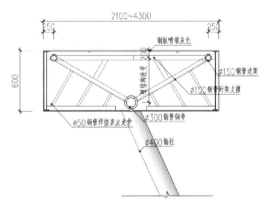

6. 构架剖面二

LANDSCAPE
DESIGN OF THE EXIT OF WUXI METRO LINE 1, ITS WIND PAVILION AND SUNKEN SQUARE

无锡地铁1号线出入口、风亭外观景观效果及下沉广场景观方案征集

Client : Railway Traffic Planning & Construction Headquater of Wuxi Municipality
Location : Wuxi
Land area : Metro line 1 with a total length of 29.42km and 24 stations
Function : landscape of traffic stops
Bidding project : Bidding project

客　　户：无锡市轨道交通规划建设领导小组（指挥部办公室）
位　　置：无锡市
用地面积：1号线全长29.42km，共设24座车站
主要功能：交通站点景观
设计竞标：设计竞标

The personality of Wuxi city can be generalized into four concepts "mountain, water, city, market ".
1. mountain----ecology and nature
2. water----water systems and the grand canal
3. city----culture and history
4. market----business and vitality

This design takes the four concepts into account, divides the twenty-four stops of line 1 into four types according to their own regional characteristics, and endows each type with unique qualities.

Corresponding with the four concepts "mountain, water, city, market ", four modern design elements are adopted in the four types of stops by means of color design, pavement design, logo design, entrance and exit design, air shaft and structure design.
1. mountain----plants; green; timber
2. water----waterscape; blue; glass
3. city----stone; orange (phonogram to the Chinese word Cheng which means city); rock
4. market----metal; red (symbolizing a warm business atmosphere); perforated aluminium, stainless steel, rust-colored weathering steel.

无锡城市个性可归纳提取为"山·水·城·市"四大特质：
1. 山·生态、自然。
2. 水·水系、运河。
3. 城·文化、历史。
4. 市·商业、活力。

本案利用这四大理念，将1号线24个站点根据各自的区域特征，分为四大类型，赋予四种不同的属性设计。

对应 "山·水·城·市"四大概念特质，寻找出四种相应的现代设计元素、材料，通过色彩设计、铺装设计、标识设计、出入口、风井及构筑物设计等等手法，分别运用到不同类别属性的站点设计中：
1. 山·植物；绿色；木材。
2. 水·水景；蓝色；玻璃。
3. 城·石头；橙色（谐音）；石材。
4. 市·金属；红色（代表热烈的商业气氛）；穿孔铝板、不锈钢、锈色耐候钢板。

无锡城市特点

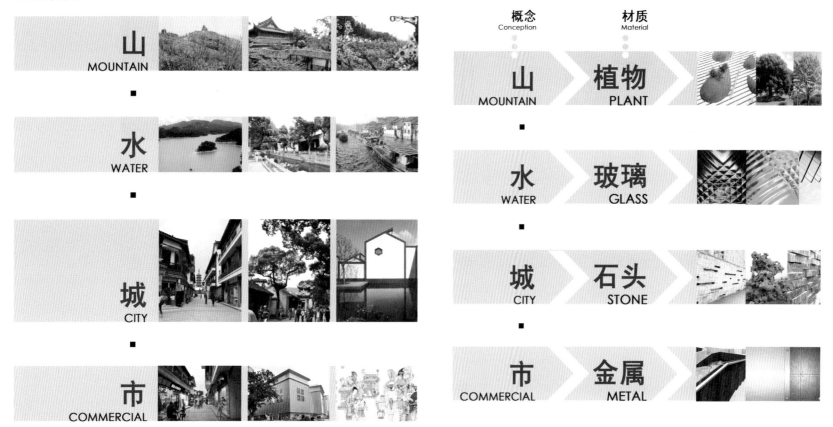

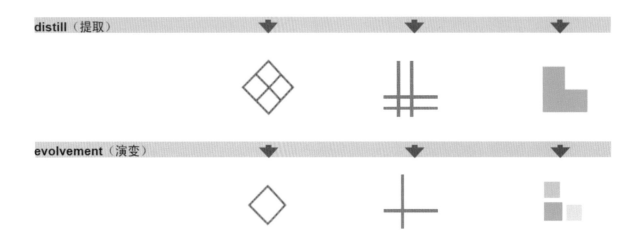

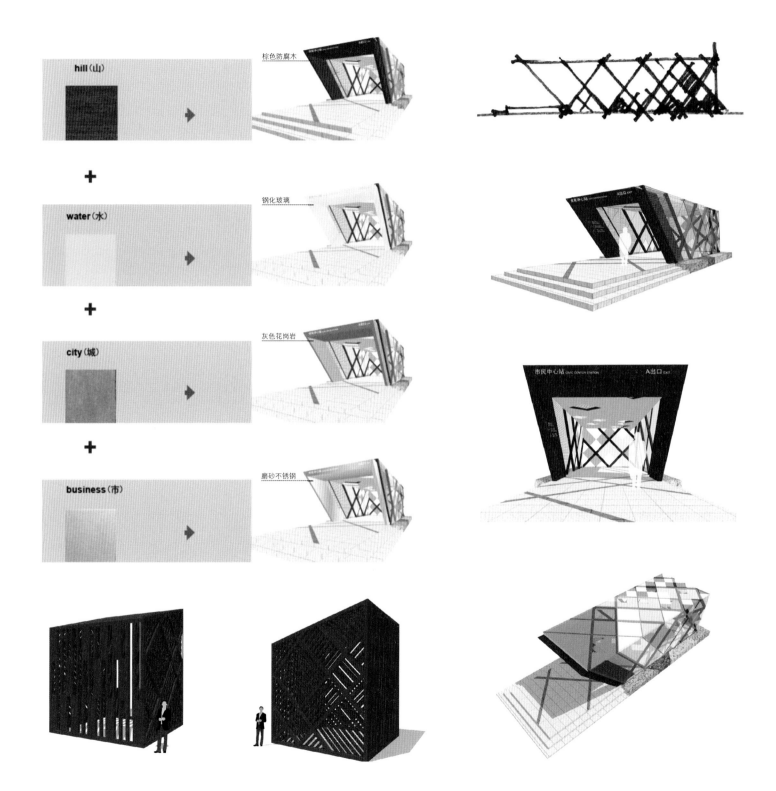

三阳广场站平面图

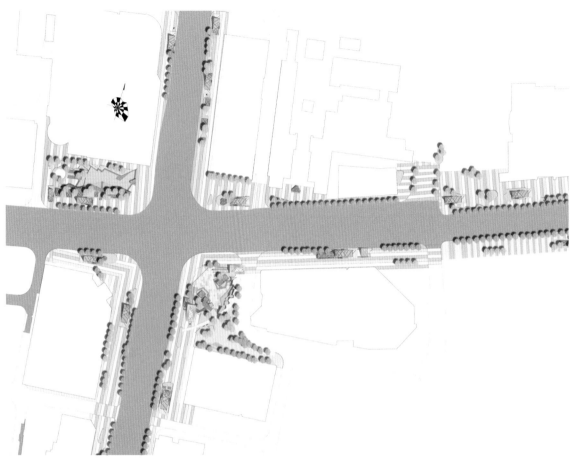

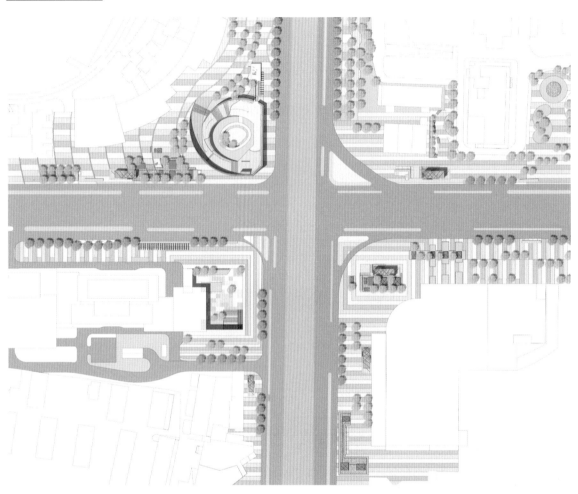

金城路站平面图

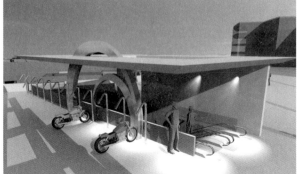

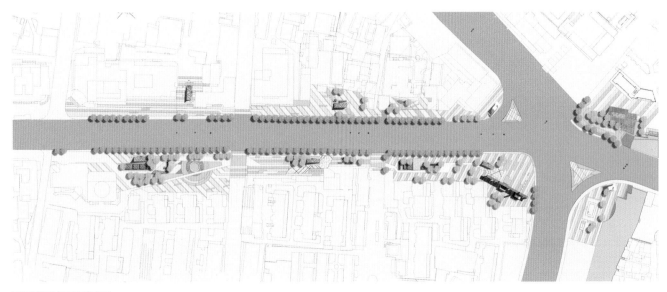

南禅寺站平面图

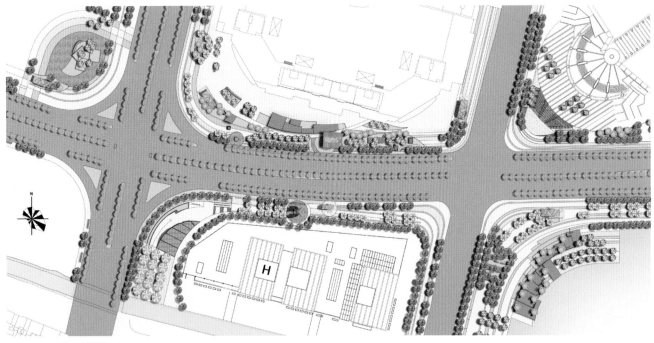

市民广场站平面图

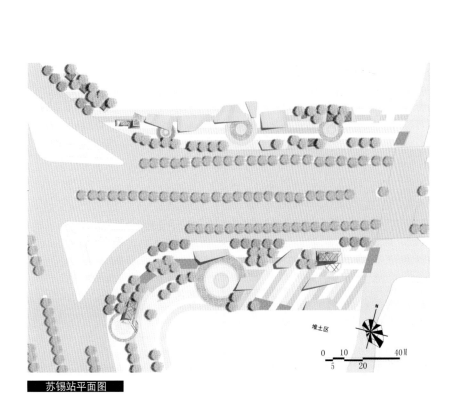

苏锡站平面图

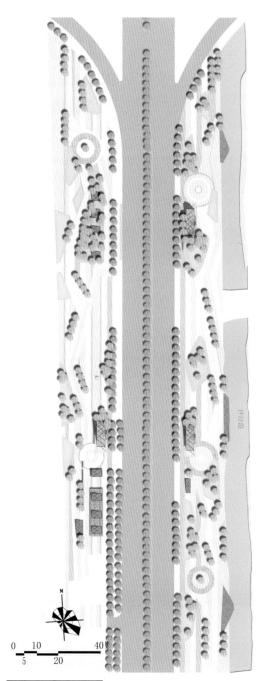

高浪站平面图

LANDSCAPE
DESIGN OF THE GROUP D OF THE RELOCATION RESIDENTIAL AREA IN ANJIN, YUAN, YUNYAN, GUIYANG

贵阳云岩渔安安井回迁安置居住区D组团景观设计

Client : Guiyang Real Estate Development Co., Ltd. of Zhongtian Urban Development Group
Location : Yunyan District, Guiyang
Land area : 9.3ha
Landscape area : 69 800m²
Function : landscape of retail & residence

客　　户：中天城投集团贵阳房地产开发有限公司
位　　置：贵阳市云岩区
用地面积：9.3hm²
景观面积：6.98万m²
主要功能：商业、住宅景观

As Guiyang is abundant in mountain forests and blessed with a unique natural eco-environment, the design endeavors to make the residential area blended in the Guiyang eco-environment.

Waves are led from rivers to the base and pavements and green belts follow the waves. Thickness varies with the altitude and colors change subtly, thus the streamlines of space and people flow accordingly.

Pixelated color blocks are adopted to connect waterscape, pavement, green belts, facilities and so on around the theme that rivers flow to the base. For example, blue blocks symbolize rivers, green ones forests, brown ones sea sand, and dark brown ones the inside of the base. And the whole pixelated design symbolizes the inherent connection between rivers and the inside of the base.

贵阳山林资源丰富，拥有得天独厚的自然生态环境，设计力求表达对贵阳生态环境的延续性演绎。

将波浪从河流引入基地，铺装与绿化跟随"波浪"的流动，在不同高差下有不同的疏密处理并具有微妙的色彩变化，空间与人的流线也随之流动起来。

运用"像素化"的色块设计，以"河水流向基地"为主题将水景、铺装、绿化、设施等关联起来。如，蓝色块水景象征河流；绿色色块象征山林；褐色色块象征海沙；深棕色色块则象征基地内部。整体的像素化设计象征了河流与基地内部的内在联系。

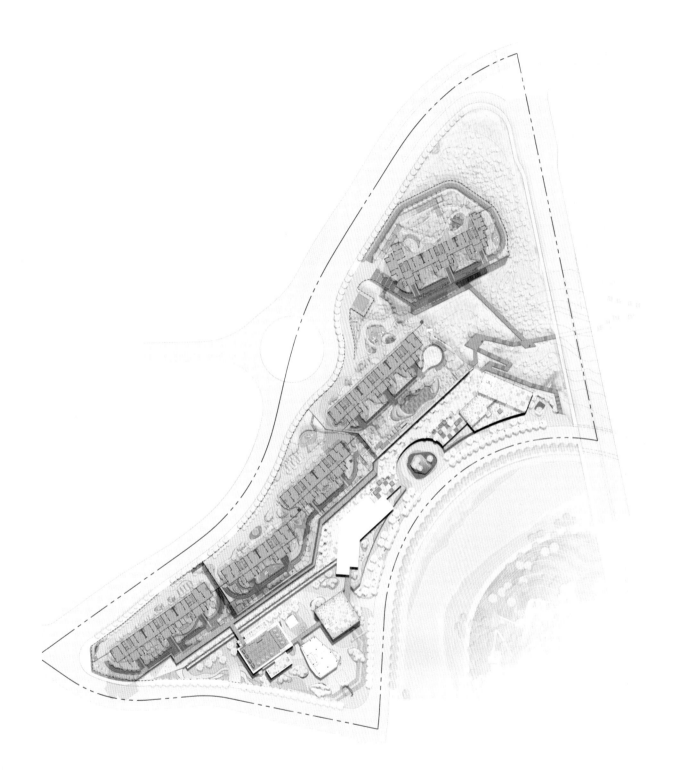

ANAALYSIS OF DISTRICT 区域分析

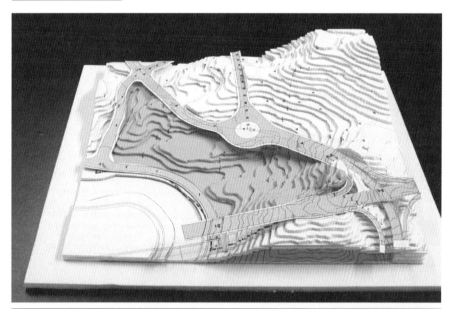

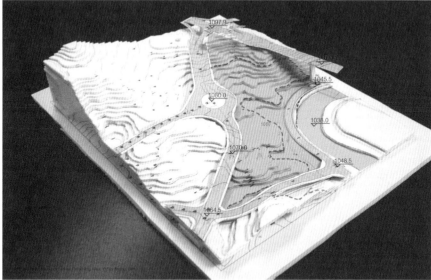

SECTION 剖面

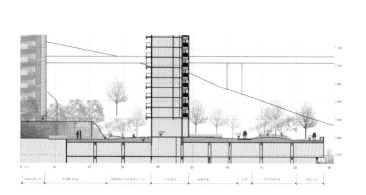

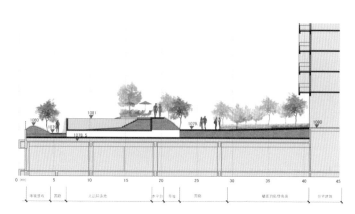

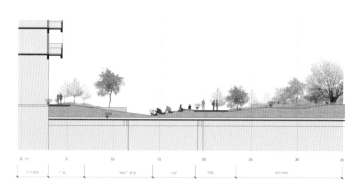

LANDSCAPE
DESIGN OF GUIYANG PLANNING EXHIBITION HALL
贵阳规划展览馆景观设计

Client : Guiyang Real Estate Development Co. Ltd of Zhongtian Urban Development Group
Location : Jinyang District, Guiyang
Land area : 3.2ha
Landscape area : 39 000m²
Function : public landscape

客　　户：中天城投集团股份有限公司
位　　置：贵阳市金阳区
用地面积：3.2hm²
景观面积：3.9万m²(含城市公共绿化带设计面积)
主要功能：公共景观

Characteristics in the plane composition and design of the project
First, the element of rectangle----the building takes the shape of a rectangle, which symbolizes the integrity and rationality of the Planning Exhibition Hall and echoes with the rectangle digital texture in the pavement of the display area.
Second, proper distribution----it results from the architectural design and the altitude differences of the base, and is also an extension of landscaping techniques in the whole area.
The whole design includes an urban square, a forest and spring square, peripheral image space, areas connecting with mountain parks and a roof garden, etc. And the roof garden tilts to the west at an angle of 15 degrees so that one can appreciate the western mountain scenery in the best way when standing in the garden and enjoys different scenes at every step in the limited space.

本项目平面构成设计中的特点
一、矩形元素----抽象于规划展览馆方正理性的矩形建筑形体，同时呼应会展区域矩形数码肌理的铺装设计。
二、错落关系----来源于建筑设计与基地高差现状，同时也是对大区域内整体造景手法的延续。
整个设计分为城市广场、林泉广场、建筑周边形象空间、与山体公园交接区域和屋顶花园几个部分。其中屋顶花园的平面采取以中心矩形西倾15°构图，通过几何变化设计形成的夹角，与西边最佳山景观赏面的方向相匹配，同时形成具有趣味的景观空间，在有限的空间里制造步移景异的游园体验。

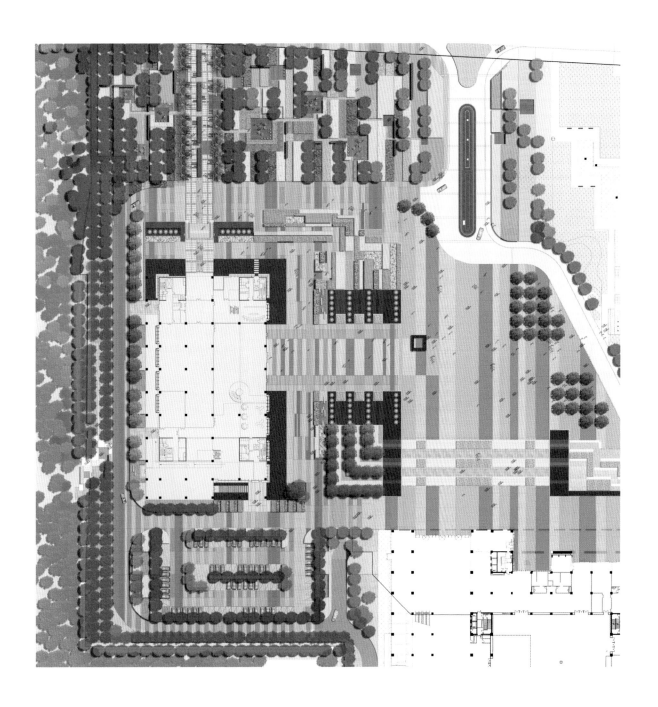

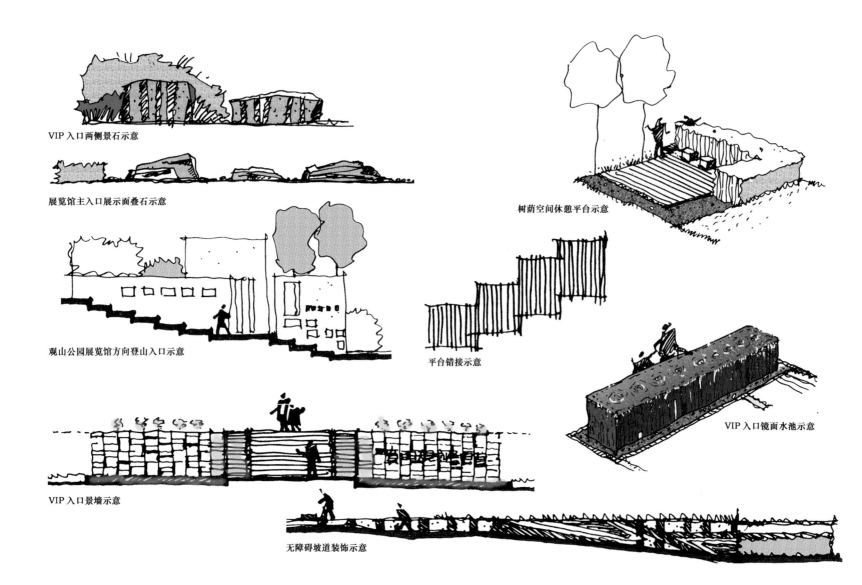

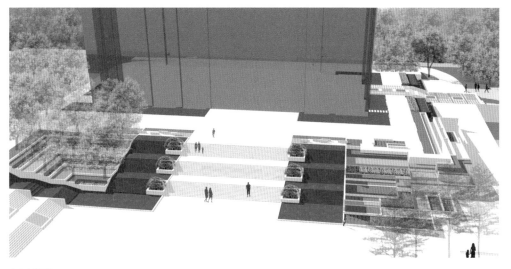

主入口透视

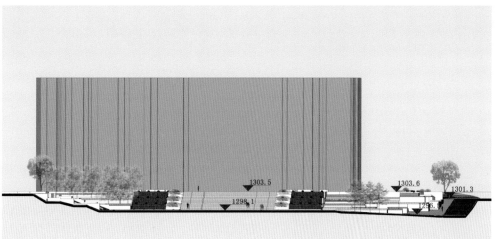

剖面 1-1

LANDSCAPE
DESIGN OF THE OSMANTHUS CENTRAL GARDEN OF THE DIANCHI LAKESIDE RESIDENCE, KUNMING
昆明滇池湖岸花园桂园中心花园

Client : Yunnan Kunchi Real Estate Co., Ltd
Location : Dianchi Road, Kunming
Land area : 3 940m²
Landscape area : 3 940m²
Function : low-rise residence
International bidding : entrusted design
under construction

客　　户：云南堃驰房地产有限公司
位　　置：昆明市滇池路
用地面积：3 940m²
景观面积：3 940m²
主要功能：低层住宅
国际竞标：委托项目
在建项目

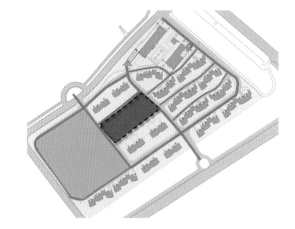

The design aims to express that the city life of quick tempo blends into natural environment through interwoven straight lines and curves. Each of the four sides of the garden is in possession of its own theme and through straight paths, all residents around the center garden can reach the center waterscape. People can enjoy varied scenes of the garden along the track. The garden can meet various needs of different people, as it includes an adventure playground surrounded by colorful flowers, an undulating garden, a tranquil bamboo grove and an open neighborhood activity square.

设计构思通过曲线与直线的穿插表达一种将快节奏的都市生活消融于曲线的自然环境中。花园四个界面各具主题，位于中心花园四周的住户都可以通过直线型的道路直接通往中心水景，同时，让游人体验不同的景观变换。场地包含一个彩花包围的儿童游乐场，地形起伏的芳香园，幽静的竹林小憩，以及开敞的邻里活动广场等，满足不同人群的多种需求。

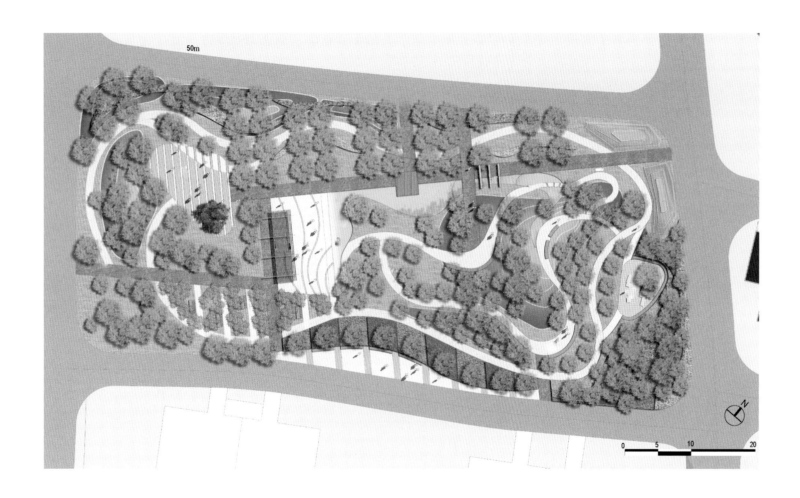

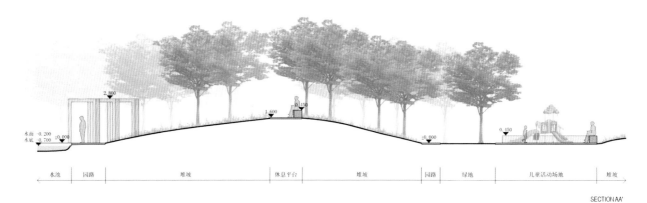

SECTION AA'

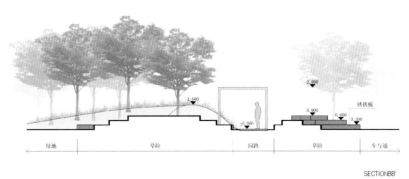

SECTION BB'

SECTION AA'

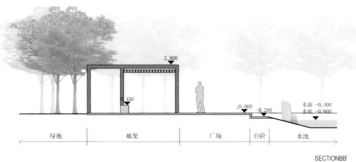

SECTION BB'

LANDSCAPE
DESIGN OF NANHAI (SOUTH SEA) MIDDLE SCHOOL, SHENZHEN
深圳南海中学景观方案设计

Client : Construction Works Bureau of Nanshan District, Shenzhen
Location : Nanshan District, Shenzhen
Land area : 2.3ha
Landscape area : 23 000m²
Function : Campus landscape

客　　户：深圳市南山区建筑工务局
位　　置：深圳市南山区
用地面积：2.3hm²
景观面积：2.3万m²
主要功能：校园景观绿化

As Nanhai Middle School is blessed with Qian Sea in the west and Nan Hill in the south, its design focuses on the theme of hill and sea. Using ellipses and curves to interpret the theme of the mutual dependence of hill and sea, the design plans to establish a flowing space so as to manifest the interactive and active teaching concepts.

The school gate is designed according to the demand that people and vehicles be separated and offers a place for waiting parents. And the perfect combination of landscape and functions makes up for deficiencies in functions of building space and endows gray space of each floor with varied landscape functions. The yard in the first floor is an exhibition of plants which shows the multiplicity of plants while serving as a place for students to carry out activities or sit and rest. The passage on the second floor is defined as a gallery for cultural display where students' compositions are displayed on non-fixed display boards in a regular way. The roof garden on the third floor is designated for the communication between teachers and students. As for roof green belts on the very top, which function mainly for energy saving and environmental protection, students can grow plants by themselves there.

南海中学西临前海，南部毗邻南山，坐拥山海环境，设计概念以山、海为主题，以椭圆和曲线的图底关系来诠释海中山，山海共生的主题，以流动而纯粹的形式串联起各个灰空间，打造连通山与海的流动空间，展现互动活泼的教学理念。

校门设计基于人车分流的功能要求设计，并提供一处供家长等候的场所。景观与功能完美结合，弥补建筑空间功能上存在的不足，赋予不同楼层的灰空间为多样的景观功能，一层建筑内庭院定义为植物展示区，既展示植物的多样性，又提供学生活动和坐憩的空间；二楼的连廊空间定义文化展示廊，以非固定展板形式来定期展示学生作品；三层屋顶花园定义为可供教师和学生交流休憩的空间；最顶层的屋顶绿化在起到节能环保的同时也提供学生一处亲手种植植物的场所。

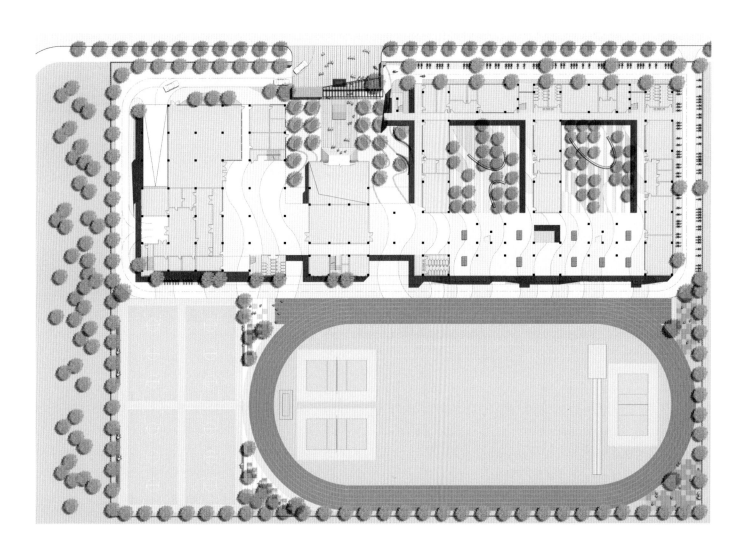

❶ 出挑花坛
❷ 移动展览板
❸ 盆栽摆放区
❹ 木座椅
❺ 特色铺地
❻ 运动场看台区

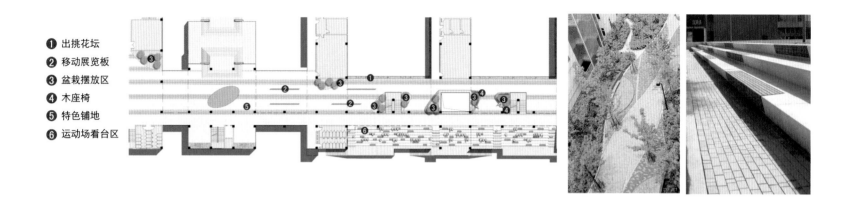

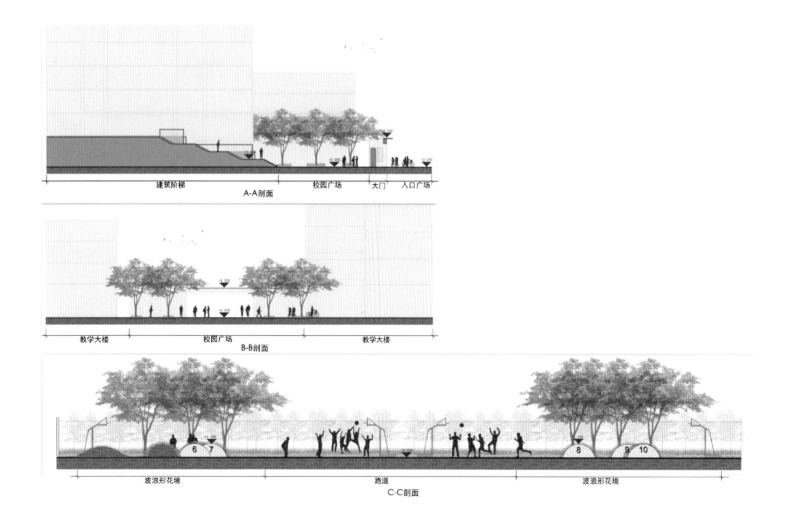

A-A剖面：建筑阶梯 | 校园广场 | 大门 | 入口广场

B-B剖面：教学大楼 | 校园广场 | 教学大楼

C-C剖面：波浪形花境 | 跑道 | 波浪形花境

LANDSCAPE
DESIGN OF THE NATIONAL MINORITIES FESTIVAL SQUARE IN GUIYANG
贵阳民族盛会广场景观方案设计

Client : Zhongtian Urban Development Group Co.,Ltd.
Location : Jinyang District, Guiyang
Land area : 7ha
Landscape area : 70 000m²
Function : Playground landscape of the National Minorities Sports Meeting 2011
under construction

客　　户：中天城投集团股份有限公司
位　　置：贵阳市金阳区
用地面积：7hm²
景观面积：7万m²
主要功能：全国少数民族运动盛会2011年会场景观
在建项目

Constructed on a site that naturally assumes the shape of a Chinese knot, the National Festival Square mainly comprises square and round in the spatial pattern, with traditional national elements integrated in a modern way. Surrounded by national totem poles, the central square is able to serve as an arena for national festivals, with various facilities constructed in accordance with the local terrain.

项目场地天然呈现"中国结"形态，以方圆成为主要空间格局。抽象及延续传统民族元素，以现代手法植入场地。中心广场满足民族庆典盛会场地需求，民族图腾柱环而置之，利用地形营造不同功能空间，以现代的表皮依附在传统的载体上。

CONCEPT 设计概念

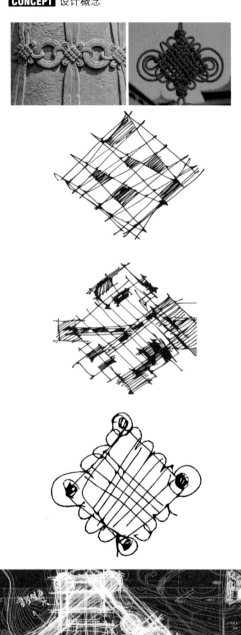
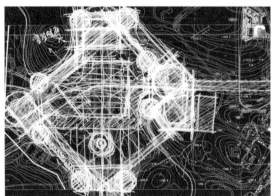
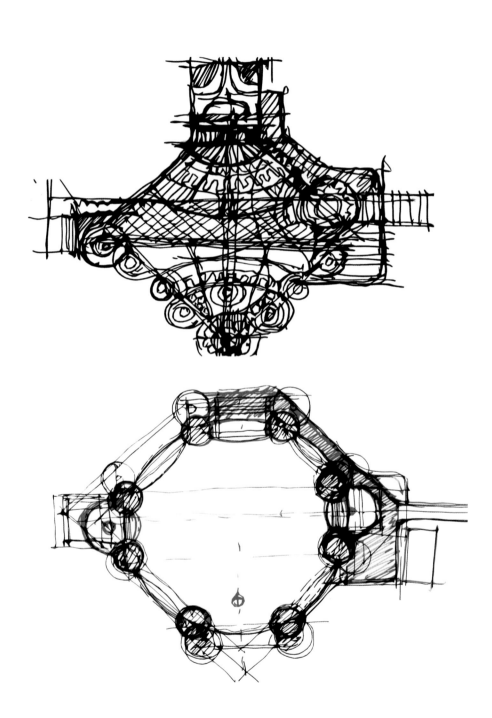

2011作品附录
2011 PROJECTS

索引

206 规划作品附录

208 建筑作品附录

212 景观作品附录

Index

206 **Appendix of planning works**

208 **Appendix of architectural works**

212 **Appendix of landscape works**

规划作品附录
Planning

项目数量: 18
已竣工或建造中项目: 4

* 已竣工或建造中项目
* Projects constructed or under construction

040

* 三亚万科湖心岛修建性详细规划及建筑方案设计
2011年，国际竞标中标项目，用地面积43.61 hm²，建筑面积45万m²
Detailed Planning and Architectural Design of Vanke Center Island Project in Sanya
2011 winning project of international bidding
Land area: 43.61 ha
Building area: 450 000 m²

048

深圳盐田翡翠岛项目规划、建筑及景观设计
2011年，国际竞标第三名，用地面积5.5 hm²，建筑面积13万m²
Planning, Architectural and Landscape Design of the Feicui/jade Island, Yantian, Shenzhen
2011 3rd prize of international bidding
Land area: 5.5 ha
Building area: 130 000 m²

054

唐山湾国际游艇社区项目概念方案规划设计
2011年，国际竞标中标方案，房地产开发部分：用地面积82.2 hm²，建筑面积56.8万m²，非房地产开发部分：用地面积75.7 hm²，建筑面积4.6万m²
Planning Design of the International Yacht Community of Tang
2011 winning project of international bidding
Development zone of real estate: Land area: 82.2 ha
Building area: 568 000 m²
Development zone of non-real estate: Land area: 75.7 ha
Building area: 46 000 m²

060

苏州太湖国家旅游度假区东入口地区城市设计
2011年，国际竞标，用地面积373 hm²，建筑面积100万m²
Urban Design of the East Entry of Taihu Lake International Tourism Zone, Suzhou
2011 international bidding project
Land area: 373 ha
Building area: 1 000 000 m²

068

深圳湾生态科技城B-TEC项目规划及建筑方案设计
2011年，国际竞标二标段入围第二名，用地面积20.32 hm²，建筑面积120万m²
Planning & Architectural Design of B-TEC City of Shenzhen Bay
2011 runner-up of Bid Section - 2 of international bidding
Land area: 20.32 ha
Building area: 1 200 000 m²

084

长城笋岗城市综合体规划及建筑概念方案设计
2011年，国际竞标，用地面积6.65 hm²，建筑面积37.1万m²
Planning & Architectural Design of the Great-w Complex Building In Sungang, Shenzhen
2011 international bidding project
Land area: 6.65 ha
Building area: 371 000 m²

088

浙江省余杭高新技术产业园区科技园一期工程规划及建筑设计
2011年，国际竞标中标方案，用地面积16.1 hm²，建筑面积39.2万m²
Planning & Architectural Design of the Yuhang High-tech Industry Park (phase I), Zhejiang
2011 winning project of international bidding.
Land area: 16.1 ha
Building area: 392 000 m²

094

杭州转塘(1-C2-03地块)项目规划及建筑设计
2011年，国际竞标中标方案，用地面积6.14 hm²，建筑面积11.1万m²
Planning & Architectural Design of the Lot 1-C2-C3 of Zhuangtang Area, Hangzhou
2011 international bidding project
Land area: 6.14 ha
Building area: 111 000 m²

福建省宁德市住宅小区建筑设计
2011年，用地面积12.28 hm²，建筑面积31.28万m²
Architectural Design of a Residential Garden Ningde, Fujian
2011 Land area: 12.28ha
Building area: 312 800 m²

...省仙桃市体育中心规划建筑设计
...年，国际竞标，用地面积21 hm²，
...面积7.4万m²

Planning & Architectural Design of Xiantao Sports
...er, Hubei Province
... international bidding project
...d area: 21 ha
...ding area: 74 000 m²

东莞海隆达兴大岭山规划建筑设计
2011年，用地面积11 hm²，建筑面积45.7万m²

Planning & Architectural Design of Dalingshan
Project developed by Hailong Daxin, Dongguan2011
Land area: 11 ha
Building area: 457 000 m²

* 沈阳万国红酒文化博览中心建筑方案设计
2011年，委托项目 用地面积9.7 hm²，建筑面积10.3万m²

Architectural Design of the Universal Exposition
Center of Wine Culture, Shenyang
2011 entrusted project　Land area: 9.7 ha
Building area: 103 000 m²

莱蒙布吉农批市场城市更新概念设计
2011年，委托项目 用地面积19.2 hm²，建筑面积37.3万m²

Urban Design of the Renovation Project of Top
Spring·Buji Agricultural Product Marketing,
Shenzhen
2011 entrusted project　Land area: 19.2 ha
Building area: 373 000 m²

...未来方舟D-1区修建性详细规划方案设计
... 委托项目 用地面积9.8 hm²，建筑面积74.42万m²

...ailed Constructive Planning of Zone D-1 of
...ngtian·Future Ark
...1 entrusted project　Land area: 9.8 ha
...ding area: 744 200 m²

* 中天·未来方舟F区修建性详细规划方案设计
2011年，委托项目 用地面积21.55 hm²，建筑面积86.7万m²

Detailed Constructive Planning of Zone F of
Zhongtian·Future Ark
2011 entrusted project　Land area: 21.55 ha
Building area: 867 000 m²

北投贵阳延安路城市综合体建筑概念性方案设计
2011年，委托项目 用地面积5.1 hm²，建筑面积34.92万m²

Architectural Design of the Complex Building on the
Yan'an Road of Guiyang
2011 entrusted project　Land area: 5.1 ha
Building area: 349 200 m²

武汉市汉阳江城大道城市设计
2011年，委托项目 用地面积85.9 hm²，建筑面积341.6万m²

Urban Design of Jiangcheng Boulevard of Hanyang,
Wuhan
2011 entrusted project　Land area: 85.9 ha
Building area: 3 416 000 m²

...利海集团贵阳扶风项目城市设计
... 委托项目 用地面积123.2 hm²，建筑面积634万m²

...an Design of Guiyang Fufeng Project of L'SEA
...ings
... entrusted project　Land area: 123.2 ha
...ding area: 6 340 000 m²

深圳航天总部基地修建性详细规划及近期建筑方案设计
2011年，国际竞标 用地面积29.9 hm²，建筑面积52.8万m²

Detailed Constructive Planning & Architectural Design
of the Headquarter Base of Shenzhen Air
2011 international bidding project
Land area: 29.9 ha
Building area: 528 000 m²

建筑作品附录
Architecture

项目数量: 30
已竣工或建造中项目: 18

* 已竣工或建造中项目
* Projects constructed or under construction

008

* 贵阳中天·未来方舟A区文化活动中心建筑及景观设计
2011年，用地面积7.2 hm²，建筑面积2.1万m²
在建项目
Architectural and Landscape Design of The Culture Center of Zhongtian · Future Ark, Guiyang
2011 Land area: 7.2 ha
Building area: 21 000 m²
under construction

018

* 贵阳中天·未来方舟样板房建筑设计
2011年，用地面积441 m²，
2011年已建成
Architectural Design of the Show Flat of Zhongtian · Future Ark, Guiyang
2011 Land area: 441 m²
Constructed on 2011

024

贵阳中天·未来方舟F3地块建筑设计
2011年，用地面积2.98 m²，建筑面积16.9万m²
Architectural Design of the Lot F3 of Zhongtian · Future Ark, Guiyang
2011 Land area: 2.98 m²
Building area: 169 000 m²

032

* 贵阳中天·未来方舟-东懋中心建筑设计
2011年，用地面积9.84 hm²，建筑面积66.74万m²
Architectural Design of the Dongmao Center of Zhongtian · Future Ark, Guiyang
2011 Land area: 9.84 ha
Building area: 667 400 m²

102

* 深圳市塘朗车辆段A区建筑概念方案设计
2011年，国际竞标第一名，用地面积4.35 hm²，
建筑面积26.1万m²
Architectural Design of the Parcel A of Tanglang Rolling Stock Depot, Shenzhen
2011 1st prize of international bidding
Land area: 4.35 ha
Building area: 261 000 m²

106

南海会馆概念性建筑设计
2011年，设计竞标，用地面积13.39 hm²，
建筑面积3.5万m²
Architectural Design of the South Sea Memorial Hall, Foshan
2011 bidding project
Land area: 13.39 ha
Building area: 35 000 m²

110

百度国际大厦建筑设计
2011年，国际竞标，用地面积1.41 hm²，
建筑面积17.12万m²
Architectural Design of Baidu International Tower
2011 international bidding project
Land area: 1.41 ha
Building area: 171 200 m²

116

* 深圳市侨城北工业区升级改造工程建筑设计
2011年，国际竞标中标方案，用地面积4.88 hm²，
建筑面积30.73万m²
Architectural Design of the Upgrading & Renewal Project of the Industrial Park in the North of Oct, Shenzhen
2011 winning project of international bidding
Land area: 4.88 ha
Building area: 307 300 m²

122

* 城建观澜仁山智水花园天际会所建筑设计
2011年，用地面积4 778 m²，建筑面积1 800 m²
在建项目
Architectural Design of the Horizon Club of Expand · The Garden of Benevolence & Wisdom, Shenzhen
2011 winning project of international biddingLand area: 4 778 m²
Building area: 1 800 m²
Under construction

126

* 中铁南方总部大厦建筑设计
2011年，国际竞标中标方案，用地面积5 253m²，
建筑面积5.81万m²
在建项目
Architectural Design of the CRSID Tower
2011 winning project of international bidding
Land area: 5 253 m²
Building area: 58 100 m²
Under construction

130

* 昆明滇池湖岸花园项目建筑设计
2011年，用地面积50.05hm²，
建筑面积15.07万m²
在建项目
Architectural Design of the Dianchi Lakeside Residence, Kunming
2011 Land area: 50.05 m²
Building area: 150 700 m²
Under construction

136

国家（青岛）通信产业园1、2、3号地块规划、建筑、景观方案设计
2011年，国际竞标第二名，用地面积4.36hm²，
建筑面积24.5万m²
Planning, Architectural & Landscape Design of the Plots 1/2/3 of the Communication Industry Park of Qingdao
2011 2ⁿᵈ prize of international bidding
Land area: 4.36ha
Building area: 245 000 m²

140

阿里巴巴深圳大厦建筑概念方案设计
2011年，国际竞标，用地面积1.38hm²，
建筑面积10.82万m²
Architectural Design of Alibaba's Branch Tower in Shenzhen
2011 International bidding project
Land area: 1.38ha
Building area: 108 200 m²

146

* 深圳市宝安区新安宝城34区改造项目建筑设计
2011年，国际竞标中标方案，用地面积1.24hm²，
建筑面积7.4万m²
Architectural Design of the Renewal Project of the 34th block of Bao'an District, Shenzhen
2011 winning project of international bidding
Land area: 1.24ha
Building area: 74 000 m²

150

深圳职业技术学院西校区综合楼工程建筑设计
2011年，设计竞标第二名，用地面积0.52 hm²，
建筑面积3.6万m²
Architectural Design of the Multiple-use Building in the West Campus of Shenzhen Polytechnic
2011 2ⁿᵈ prize of bidding
Land area: 0.52 ha
Building area: 36 000 m²

* 贵阳中天世纪新城九 十组团建筑概方案设计
2011年，设计竞标，用地面积1.9 hm²，
建筑面积9.8万m²
Architecture Design of the Nine&Ten Group of the Zhongtian Century City, Guiyang
2011 Bidding project
Land area: 1.9 ha
Building area: 98 000 m²

深圳宝安龙圣堡一期建筑设计
2011年，设计竞标，用地面积3.7 hm²，
建筑面积14.9万m²
Architecture Design of Longshengbao Residence Garden (phase I) in Baoan District, Shenzhen
2011 Bidding project
Land area: 3.7 ha
Building area: 149 000 m²

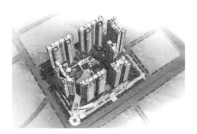

华联宝安27区项目建筑方案设计
2011年，设计竞标，用地面积3.75hm²，
建筑面积27.62万m²
Architectural Design of the Hualian Residential Project in 27th Block of Bao'an, Shenzhen
2011 bidding project
Land area: 3.75ha
Building area: 276 200 m²

深圳市高新技术企业联合总部大厦建筑设计
2011年，国际竞标，用地面积0.53hm²，
建筑面积11.39万m²
Architectural Design of the United Headquarter of High-Tech Enterprises of Shenzhen
2011 international bidding project
Land area: 0.53 ha
Building area: 113 900m²

清华大学深圳研究生院创新基地建筑设计
2011年，国际竞标，用地面积0.96hm²，
建筑面积6.1万m²
Architectural Design of the Innovation Base of Tsinghua University Graduate School at Shenzhen
2011 international bidding project
Land area: 0.96ha
Building area: 61 000 m²

佛山高铁西站建筑方案设计
2011年，设计竞标，用地面积80hm²，
建筑面积10.4万m²
Architectural Design of Foshan West Station
2011 international bidding project
Land area: 80ha
Building area: 104 000m²

* 鹏瑞中心·深圳湾1号北区建筑设计
2011年，委托项目 用地面积1.37 hm²，
建筑面积11.9万m²
Architecture Design Pan One Center·Shenzhen Bay NO.1
2011 entrusted project Land area: 1.37 ha
Building area: 119 000 m²

福建三环路洪塘段安置房建筑方案设计
2011年，国际竞标 用地面积6.39 hm²，
建筑面积25.02万m²
Architectural Design of Hongtang Resettlement House on Sanhuan Road, Fujian
2011 International bidding project
Land area: 6.39 ha Building area: 250 200 m²

* 深业·巴黎宫苑建筑方案设计
2011年，委托项目 用地面积13.57 hm²，
建筑面积3.8万m²
Architectural Design of SumYep · Paris Palace
2011 entrusted project
Land area: 13.57 ha Building area: 38 000 m²

* 中航惠州巽寮湾住宅建筑设计
2011年，委托项目 用地面积32 hm²，
建筑面积56万m²
Architectural Design of AVIC Xuliao Residential Project, Huizhou
2011 entrusted project
Land area:32 ha Building area: 560 000 m²

* 贵阳中天·未来方舟F1地块建筑设计
2011年，委托项目
用地面积1.93 hm²，建筑面积12.1万m²
Architectural Design of the Plot F1 of Zhongtian·Future Ark,Guiyang
2011 entrusted project
Land area: 1.93 ha Building area: 121 000 m²

* 贵阳中天·未来方舟F4地块建筑设计
2011年，委托项目
用地面积4.43 hm²，建筑面积31.9万m²
Architectural Design of the Plot F4 of Zhongtian·Future Ark,Guiyang
2011 entrusted project
Land area: 4.43 ha Building area: 319 000 m²

* 贵阳中天·未来方舟F6地块建筑设计
2011年，委托项目
用地面积1.32 hm²，建筑面积11.01万m²
Architectural Design of the Plot F6 of Zhongtian·Future Ark,Guiyang
2011 entrusted project
Land area: 1.32 ha Building area: 110 100 m²

* 贵阳中天·未来方舟F7地块建筑设计
2011年，委托项目
用地面积2.35 hm²，建筑面积14.85万m²
Architectural Design of the Plot F7 of Zhongtian·Future Ark,Guiyang
2011 entrusted project
Land area: 2.35 ha Building area: 148 500 m²

* 贵阳中天·未来方舟F9地块建筑设计
2011年，委托项目
用地面积2.44 hm²，建筑面积11.58万m²
Architectural Design of the Plot F9 of Zhongtian·Future Ark,Guiyang
2011 entrusted project
Land area: 2.44 ha Building area: 115 800 m²

景观作品附录
Landscape

项目数量：10
已竣工或建造中项目：7

* 已竣工或建造中项目
* Projects constructed or under construction

158

无锡地铁1号线胜利门站广场景观方案征集
2011年，设计竞标，用地面积 2hm²，景观面积2万m²
Landscape Design of the Victory Gate Station Square of Wuxi Metro Line 1
2011 bidding project
Land area: 2 ha
Landscape area: 20 000 m²

168

无锡地铁1号线出入口、风亭外观景观效果及下沉广场景观方案征集
2011年，设计竞标，用地面积 1号线全长29.42km，共设24座车站
Landscape Design of The Exit of Wuxi Metro Line 1, Its Wind Pavilion and Sunken Square
2011 bidding project
Land area: subway line 1 with a total length of 29.42km and 24 stations

178

* 贵阳云岩渔安安井回迁安置居住区D组团景观设计
2011年，用地面积9.3 hm²，景观面积6.98万m²
Landscape Design of the Group D of the Relocation Residential Area in Anjin, Yuan, Yunyan, Guiyang
2011. Land area: 9.3 ha
Landscape area: 69 800 m²

186
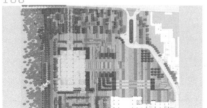

* 贵阳规划展览馆景观设计
2011年，用地面积3.2hm²，景观面积3.9万m²
Landscape Design of Guiyang Planning Exhibition Hall
2011 bidding project
Land area: 3.2ha
Landscape area: 39 000 m²

190

* 昆明滇池湖岸花园桂园中心花园
2011年，设计委托，用地面积 3 940 m²，景观面积3,940m²
在建项目
Landscape Design of the Osmanthus Central Garden of the Dianchi Lakeside Residence, Kunming
2011 entrusted design
Land area: 3 940 m²
Landscape area: 3 940 m²
under construction

196
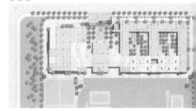

深圳南海中学景观方案设计
2011年，用地面积2.3hm²，景观面积2.3万m²
Landscape Design of Nanhai (South Sea) Middle School, Shenzhen
2011. Land area: 2.3 ha
Landscape area: 23 000 m²

200

* 贵阳民族盛会广场景观方案设计
2011年，用地面积7 hm²，景观面积7万m²
在建项目
Landscape Design of the National Minorities Festival Square in Guiyang
2011. Land area:7ha
Landscape area: 70 000 m²
under construction

* 昆明滇池湖岸花园项目住宅及配套部分景观设计
2011年，委托项目，用地面积50.05hm²，景观面积28.67万m²
在建项目
Landscape Design of the National Minorities Festival Square in Guiyang
2011 entrusted project
Land area: 50.05ha
Landscape area: 286 700 m²
under construction

* 南山商业文化中心二层连廊桥底景观设计
2011年，委托项目，用地面积1.6hm²，景观面积1.6万m²
Landscape Design Under the Air Corridor, Nanshan Commercial and Cultural Center
2011 entrusted project
Land area:1.6ha
Landscape area: 16 000 m²

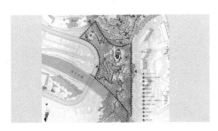

中天·未来方舟活动中心周边山体公园景观设计
2011年，委托项目　用地面积1.89 hm²，
景观面积1.89万m²
Landscape Design of the Mountain Park Surrounding
Zhongtian Future Ark Activity Center
2011 Mandated project Land area:1.89 ha
Landscape area: 18 900 m²

往年作品附录
ANNEX

索引

216 规划

226 公建

234 住宅

240 景观

Index

216 **Planning**

226 **Public Building**

234 **Residence**

240 **Landscape**

规划
Planning

项目总数: 156
已竣工或建造中项目: 24

* 已竣工或建造中项
* Projects constructed or under construction

* 贵阳云岩区渔安安井片区城市设计
2010年，委托设计，用地面积755 hm²，建筑面积650万m²
Planning and Architecture Design of Yu'an & An Jing of Yunyan District in Guiyang
2010 Mandated project Land area: 755 ha
Building area: 6 500 000 m²

武汉新区四新生态新城"方岛"区域城市设计
2010年，国际竞标第二名，用地面积268 hm²，建筑面积400万m²
Urban Design of "Square Island" of Six Ecological Town in the New Area of Wuh
2010 Second place of international bidd
Land area: 268 ha Building area: 4 000 m²

台湾高雄海洋文化及流行音乐中心新建工程
国际竞赛
2010年，国际竞标，用地面积10 hm²，建筑面积7.1万m²
Architecture Design of Kaohsiung Maritime Culture & Popular Music Center
2010 International bidding Land area: 10 ha
Building area: 71 000 m²

* 佛山西站综合交通枢纽概念性规划建筑设计
2010年，国际竞标中标方案，用地面积约31 hm²
建筑面积约8万m²
Planning and Architecture Design of Foshan West Station
2010 Winning project of international bidding
Land area: 31 ha Building area: 80 000 m²

长春南部新城净月西区生态商务金融中心（EBD）城市设计
2010年，国际竞标第二名，用地面积319.5 hm²，
建筑面积574.5万m²
Urban Design of EBD of Jingyue District, Changchun
2010 Second place of international bidding
Land area: 319.5 ha Building area: 5 745 000 m²
Building area: 5 745 000 m²

花溪山水度假旅游城贵阳花溪概念性城市设计
2010年，委托设计，用地面积108.5 km²，
建筑面积5800万m²
Conceptual Urban Design of Huaxi Touris Resort of Guiayang
2010 Mandated project Land area: 108 km² Building area: 58 000 000 m²

中粮地产（集团）深圳宝安61区概念性规划建筑设计
2010年，国际竞标，用地面积 2.4 hm²，
建筑面积9.8万m²
Conceptual Planning and Architecture Design of Zone 61 of COFCO Real Estate, Bao'an District, Shenzhen
2010 International bidding Land area: 2.4 ha
Building area: 98 000 m²

贵阳星云家电城项目城市设计及单体建筑概念性方案设计
2010年，委托设计，用地面积 3.25 hm²
建筑面积52.7万m²
Conceptual Urban Planning & Architecture Design of Xingyun Home Appliances City of Guiyang
2010 Mandated project Land area: 3.25 ha
Building area: 527 000 m²

航天成都城上城规划、建筑设计
2010年，国际竞标中标方案，用地面积 4.27hm²，
建筑面积30万m²，高度70m
Planning and Architecture Design of Aerospace Town in Chengdu
2010 Winning project of international bidding Land area: 4.27 ha
Building area: 300 000 m² Height: 70 m

* 深圳半岛城邦三四五期详细蓝图修编
2010年，委托设计，用地面积 26.42 hm²，建筑面积93万m²，建筑高度100~200m
Detailed Blueprint Modification of Peninsula Residential Community in Shenzhen
2010 Mandated project Land area:26.42 ha
Building area: 930 000 m² Height: 100~200m

* 福州中央商务中心安置房规划建筑设计
2010年，中标方案，用地面积 4.75 hm²，
建筑面积23万m²，建筑高度<100m
Planning and Architecture Design of Resettlement Housing of Fuzhou CBD
2010 Winning project Land area: 4.75 ha
Building area: 230 000 m² Height: <100 m

贵阳市河滨剧场片区旧城改造项目
2010年，委托设计，用地面积13hm²，
建筑面积100万m²
Urban Renewal of Waterfront Theatre Area of Guiyang
2010 Mandated project Land area: 13ha
Building area: 1 000 000 m²

贵阳市人民剧场片区旧城改造项目
2010年，委托设计，用地面积5.7hm²，
建筑面积60万m²
Urban Renewal of The People Theatre Area of Guiyang
2010 Mandated project Land area: 5.7ha
Building area: 600 000 m²

马家龙工业区04-05-17地块改造设计
2010年，委托设计，用地面积1.5hm²，
建筑面积7.9万m²
Urban & Archiectural Design for the Reconstruction of Plot 04-05-17 of Majialong Industrial Zone, Shenzhen
2010 Mandated project Land area: 1.5ha
Building area: 79 000 m²

* 贵阳渔安安井片区原回迁区D组团地块修建性详细规划
2010年，委托设计，用地面积9.2hm²，
建筑面积22万m²
Constructive Detailed Planning of Bloc D of Anjing, Yuan District, Guiyang
2010 Mandated project Land area: 9.2ha
Building area: 220 000 m²

* 贵阳国际会议展览中心修建性详细规划
2009年，委托设计，用地面积51.8hm²，
建筑面积88.88万m²
Construction planning of Guiyang International Conference & Exposition Center
2009 Mandated project Land area: 51.8 ha
Building area: 888800m²

* 中山金源花园规划设计
2009年，中标方案，用地面积21.22hm²，
建筑面积79.26万m²
Planning & Architectural Design of Jin Yuan Garden Residence in Zhongshan
2009 Winning project Land area: 21.22ha
Building area: 792600m²

南方科技大学和深圳大学新校区拆迁安置项目·商住综合区设计
2009年，竞标方案，用地面积17.77hm²，建筑面积60万m²
Removing and Resettlement of New Campus of South University of Science and Technology and Shenzhen University - Planning & Architectural Design for Commerce and Residence
2009 Bidding project Land area: 17.77ha
Building area: 600000m²

河源万绿湖风景区首期规划设计
2009年，委托设计，用地面积32hm²
Planning Design of Wanlv Lake Holiday and Tourism Scenic Area (Phase II), Heyuan
2009 Mandated project Land area: 32ha

安徽铜陵一号地块概念规划
2009年，委托设计，用地面积27.7hm²，
建筑面积25万m²
Conceptional Planning of 1# Auction Site of Tongling, Anhui
2009 Mandated project Land area: 27.7ha
Building area: 250000m²

深圳光明新区文化艺术中心及光明医院规划研究
2009年，委托设计，用地面积29.6hm²
Planning Research for the Sites of Center and Hospital of Guangming New District
2009 Mandated project Land area: 29.6ha

深圳远致创业园规划设计
2009年，竞标方案，用地面积12.17hm²，
建筑面积60.67万m²
Planning & Architectural Design of Yuanzhi Innovation Park
2009 Bidding project Land area: 12.17ha
Building area: 606,700m²

深圳观澜横坑招拍卖地块前期概念规划
2009年，委托设计，用地面积15.78hm²，
建筑面积28.57万m²
Conceptional Planning of the Auction Site in Hengkeng of Guanlan, Shenzhen
2009 Mandated project Land area: 15.78ha
Building area: 285700m²

* 贵阳云岩区渔安安井片区规划设计
2009年，委托设计，用地面积1200hm²，
建筑面积450万m²
Planning Design of Yu'an & An Jing of District, Guiyang
2009 Mandated project Land area: 1200ha
Building area: 4500000m²

金地东莞黄江居住社区概念性规划设计
2009年，国际竞标，用地面积25hm²，
建筑面积28万m²
Planning Design of Gemdale Huangjiang Residential Community
2009 International bidding Land area: 25ha
Building area: 280000m²

深圳光明新区招拍卖地块前期概念规划
2009年，委托设计，用地面积9.07hm²，
建筑面积18.15万m²
Conceptional Planning of the Auction Site of Guangming New District, Shenzhen
2009 Mandated project Land area: 9.07ha
Building area: 181500m²

深圳市后海中心区东滨路项目概念方案设计
2009年，委托设计，用地面积4.6hm²，
建筑面积35万m²
Conceptional Urban & Architectural
Design for Dongbin Road Project, Shenzhen
2009 Mandated project Land area: 4.6ha
Building area: 350000m²

贵阳新火车站片区及商业金融区城市设计
2009年，委托设计，用地面积877.16hm²，
建筑面积1600万m²
Urban Design of Guiyang New Railway
Station Area & Commercial and Financial Area
2009 Mandated project Land area: 877.16ha
Building area: 16000000m²

惠州仲恺高新区总部经济区规划设计
2009年，竞标方案，用地面积5.42hm²，
建筑面积17.55万m²
Planning & Architectural Design of Headquarters
Economic Zone of Huizhou Zhongkai
High-Tech Industrial Park
2009 Bidding project Land area: 5.42ha
Building area: 175500m²

深圳机场开发区西区"天空之城"概念性规划设计
2009年，中标方案，用地面积55hm²，
建筑面积130万m²
Conceptional Planning Design of Sky City
(west area of Shenzhen Airport
Development Zone), Shenzhen
2009 Winning project Land area: 55ha
Building area: 1300000m²

深圳中粮宝安61区概念性规划设计
2009年，委托设计，用地面积2.43hm²，建筑面积6.79万m²
Conceptional Planning Design of Zone 61 of
Bao'an District, Shenzhen
2009 Mandated project Land area: 2.43ha
Building area: 679000m²

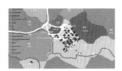

深圳南澳鹅公村改造规划设计
2009年，竞标方案，用地面积10hm²
Reconstruction Planning & Conceptional
Architectural Design for E Gong Village, Shenzhen
2009 Bidding project Land area: 10ha

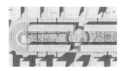

深圳高新园区软件产业基地规划设计（第三、四标段）
2009年，竞标方案，用地面积14.9hm²，
建筑面积61万m²
Planning & Architectural Design of Software
Industry Base (section III & IV) of Shenzhen
High-Tech Industrial Park
2009 Bidding project Land area: 14.9ha
Building area: 610000m²

泉州洛江现代世界农业生态休闲观光园概念规划
2009年，委托设计，用地面积510.8hm²，
建筑面积76.42万m²
Concept Planning of Ecological Town
for Luojiang Modern Agricultural Touris
Quanzhou
2009 Mandated project Land area: 510
Building area: 764200m²

南方科技大学和深圳大学新校区拆迁安置项目-产业园区规划设计
2009年，竞标方案，用地面积15.25hm²，建筑面积63.10万m²
Removing and Resettlement of New Campus of South
University of Science and Technology and Shenzhen
University-Planning & Architectural Design
of Industry Park
2009 Bidding project Land area: 15.25ha
Building area: 631000m²

深圳市光明新区保障性住房规划设计
2008年，用地面积4.12hm²，
建筑面积13万m²
Detailed Planning of Social Community of
Guangming New Town, Shenzhen
2008 Land area: 4.12ha
Building area: 130000m²

昆明尚居五甲塘概念性总体规划
2008年，用地面积62.52hm²，
建筑面积57.34万m²
Conceptual Planning of
Wujiatang Community of S-Home, Kunming
2009 Land area: 65.52ha
Building area: 573400m²

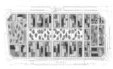

深圳赛格日立工业区升级改造城市设计
2008年，用地面积13hm²
Urban Design of Renewal of
Seg Hitachi Industry Zone in Shenzhen
2008 Land area: 13ha

深圳人才园规划设计
2008年，用地面积3.61hm²，
建筑面积8.32万m²
Detailed Planning of
Human Resources Park, Shenzhen
2008 Land area: 3.61ha
Building area: 83200m²

珠海歌剧院规划设计
2008年，用地面积42hm²，
建筑面积4.3万m²
Urban Design of Zhuhai Opera House
2008 Land area: 427ha
Building area: 43000m²

成都怡湖玫瑰湾详细规划设计
2008年，用地面积12.57hm²，
建筑面积62.5万m²
Detailed Planning of Yihu Rose Bay
Community, Chengdu
2008 Land area: 12.57ha
Building area: 625000m²

深圳市光明新区科技公园周边地区整体城市设计行政中心详细城市设计
2008年，用地面积520hm²
Urban Design of Science & Technology
Park and Administration Center of
Guangming New Town, Shenzhen
2008 Land area: 520ha

长春市高新区C-6地块住宅概念规划设计
2008年，用地面积8.57hm²，
建筑面积15.97万m²
Detailed Planning of Residential Community of
High-Tech C-6 Block, Changchun
2008 Land area: 8.57ha
Building area: 159701m²

深圳市光明新区公明文化艺术和体育中心规划设计
2008年，用地面积9.9hm²，
建筑面积5.09万m²
Urban Design of Gongming Culture, Art & Sports
Center of Guangming New Town, Shenzhen
2008 Land area: 9.9ha
Building area: 50900m²

深圳龙岗坪山街道宝山第二工业区改造概念性规划设计
2008年，用地面积74.92hm²，
建筑面积225.72万m²
Urban Design of Renewal of Baoshan 2nd Industry
Zone in Pinshan of Longgang District, Shenzhen
2008 Land area: 74.92ha
Building area: 2257200m²

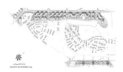

贵州遵义商业街项目概念规划设计
2008年，用地面积11.69hm²，
建筑面积60万m²
Conceptual Planning of Zunyi
Commercial Street, Guizhou
2008 Land area: 11.69ha
Building area: 600000m²

深圳天鹅堡三期概念规划设计
2008年，用地面积11.17hm²，
建筑面积20万m²
Conceptual Planning & Design of
OCT Swan Castle (Phase3), Shenzhen
2008 Land area: 11.17ha
Building area: 200000m²

宿州综合体概念性规划设计
2008年，用地面积21.3hm²，
建筑面积67.92万m²
Conceptual Planning of Suzhou Complex
2008 Land area: 21.3ha
Building area: 679200m²

长春市净月区梧桐街住宅规划设计
2008年，用地面积34.8hm²，
建筑面积45万m²
Planning Design of Residential Community of
Phoenix Tree Street of Jingyue District in Changchun
2008 Land area: 34.8ha
Building area: 450000m²

中国饮食文化城城市设计
2008年，用地面积220hm²，
建筑面积38万m²
Urban Design of Chinese Gastrologic and
Cultural Town in Shenzhen
2008 Land area: 220ha
Building area: 380000m²

深圳市南油购物公园城市设计
2008年，国际竞标第一名，
用地面积13hm²，建筑面积45.4万m²
Urban Design of
Nanyou Shopping Park, Shenzhen
2008 First prize of international bidding

深圳光明新城万丈坡居住小区详细蓝图设计
2008年，详蓝编制范围22hm²，规划研究范围49.6hm²，
建筑面积53.23万m²
Detailed Blueprint Design of Wanzhangpo Residential
Community of Guangmin New Town, Shenzhen
2008 Scope of detailed blueprint: 22ha
Scope of planning research: 49.6ha
Building area: 532,300m²

深圳南澳月亮湾海岸带景观改造设计概念规划
2008年，用地面积14.2hm²
Conceptual Plannning Design for Landscape
Reconstruction of Coastal Zones of Nan'ao, Shenzhen
2008 Land area: 14.2ha

深圳市光明新区中央公园概念规划方案
2008年，用地面积237hm²
Conceptual Planning Design of Central
Park of Guangming District, Shenzhen
2008 Land area: 237ha

珠海五洲花城二期概念性规划设计
2008年，用地面积16.5hm²，
建筑面积60万m²
Conceptual Planning of Five Continental
Residence Garden, Zhuhai
2008 Land area: 16.5ha
Building area: 600000m²

朗钜呼和浩特高尔夫社区概念规划
2008年，用地面积282hm²，
建筑面积150万m²
Conceptual Planning of
Large's Golf Residence Community, Hohhot
2008 Land area: 282ha
Building area: 1500000m²

贵阳中天集团中华北路片区概念规划设计
2008年，用地面积32.67hm²，
建筑面积232.16万m²
Conceptual Planning of
Zhonghua North Street Block of
Zhongtian Group, Guiyang
2008 Land area: 32.67ha
Building area: 2321600m²

中山市东凤镇新沙岛规划设计
2008年，用地面积80.5hm²，
建筑面积80万m²
Urban Design of Xinsha Island of
Dong feng Town, Zhongshan
2008 Land area: 80.5ha
Building area: 800000m²

长沙华润含浦项目概念性规划设计
2008年，用地面积15hm²，
建筑面积45万m²
Conceptual Planning of CRL Hanpu Residential
Community, Changsha
2008 Land area: 15ha
Building area: 450000m²

成都天府华侨城二期概念规划设计
2008年，用地面积10hm²，
建筑面积15万m²
Conceptual Planning of
Tianfu OCT Residence (Phase 2), Chengdu
2008 Land area: 10ha
Building area: 150000m²

中山市坦州镇居住小区规划设计
2008年，用地面积11.4hm²，
建筑面积28万m²
Detailed Planning of Tanzhou Residential
Community, Zhongshan
2008 Land area: 11.4ha
Building area: 280000m²

★ 昆明滇池旅游度假区文化公园规划设计
2007年，用地面积111hm²，
建筑面积88.8万m²
Planning Design of Cultural Park of
Dianchi Tourism & Resort Region, Kunming
2007 Land area: 111ha
Building area: 888000m²

★ 惠州鹏基半山名苑规划设计
2007年，用地面积49.71hm²，
建筑面积64.1万m²，竣工日期：2009年
Planning Design of Pengji Hillside Residential
Community, Huizhou
2007 Land area: 49.71ha
Building area: 641000m² Constructed in 2009

江苏省姜堰市锦绣城规划设计
2007年，用地面积29.62hm²，
建筑面积62万m²
Planning Design of Jiangyan Splendid Town,
Jiangsu
2007 Land area: 29.62ha
Building area: 620000m²

★ 深圳城建·观澜居住区规划设计
2007年，用地面积16.99hm²，
建筑面积34.6万m²
Planning Design of Guanlan Residential
Community, Shenzhen
2007 Land area: 16.99ha
Building area: 346000m²

深圳中兴大梅沙培训基地规划设计
2007年，用地面积11hm²，
建筑面积11.94万m²
Planning Design of ZTE Formation
Base in Dameisha Shenzhen
2007 Land area: 11ha
Building area: 119400m²

清远新时代广场项目概念规划方案设计
2007年，用地面积55.19hm²，
建筑面积82.37万m²
Conceptual Planning Design of
New Times Plaza, Qingyuan
2007　Land area: 55.19ha
Building area: 823700m²

江苏连云港新华社区概念规划研究
2007年，用地面积97hm²，
建筑面积101.13万m²
Conceptual Planning Research of
Xinhua Residential Community, Lianyungang, Jiangsu
2007　Land area: 97ha
Building area: 1011300m²

东莞黄江伯爵山庄概念性规划设计
2007年，用地面积103.47hm²，
建筑面积36万m²
Planning Design of Earl Resort Hotel, Dongguan
2007　Land area: 103.47ha
Building area: 360000m²

深圳市观澜版画基地概念性规划
2007年，用地面积150hm²，
建筑面积19万m²
Planning Design of Guanlan Art Print Base,
Bao'an district, Shenzhen
2007　Land area: 150ha
Building area: 190000m²

* 成都华润置地翡翠城小学规划设计
2007年，用地面积1.65hm²，
建筑面积1.71万m²
Planning Design of
CRL Jade-City Primary School, Chengdu
2007　Land area: 1.65ha
Building area: 17100m²

深圳市罗湖区贝丽中学（水贝珠宝学校）规划设计
2007年，国际竞标，用地面积1.95hm²，
建筑面积2.53万m²
Planning Design of Shenzhen Beili Middle School
2007　International bidding
Land area: 1.95ha
Building area: 25300m²

* 深圳市龙岗区天安数码新城二期规划设计
2007年，国际竞标中标方案，用地面积4.63hm²，
建筑面积24.68万m²
Planning Design of Longgang Cyber Park, Shenzhen
2007　Winning project of international bidding
Land area: 4.63ha
Building area: 246800m²

佛山南海区狮山镇客运站修建性详细规划
2007年，国际竞标中标方案，用地面积10hm²，
建筑面积13.14万m²
Detailed Planning of Shishan
Long Distance Bus Station, Foshan
2007　Winning project of international bidding
Land area: 10ha
Building area: 131400m²

佛山南海区狮山路镇文化体育公园规划设计
2007年，用地面积13.90hm²，
建筑面积28万m²
Planning Design of Shishan Cultural & Sport Park,
Nanshan, Foshan
2007　Land area: 13.90ha
Building area: 280000m²

珠海前山新冲城市壹站规划设计
2007年，国际竞标，用地面积2.85hm²，
建筑面积8.31万m²
Planning Design of the First City Station
Residential Community, Zhuhai
2007　International bidding
Land area: 2.85ha
Building area: 83100m²

深圳市龙岗区"深业·坪山"居住区规划设计
2007年，国际竞标，用地面积2.83hm²，
建筑面积11.7万m²
Planning Design of 'Shum Yip Pingshan'
Residential Community, Shenzhen
2007　International bidding
Land area: 2.83ha
Building area: 117000m²

苏州太湖旅游度假区东入口区域概念性规划设计
2007年，中标方案，用地面积43hm²
Conceptual Planning of East Entrance Area,
Taihu Lake National Holiday zone, Suzhou
2007　Winning project　Land area: 43ha

* 贵阳中天世纪新城三号地块六、七组团规划设计
2007年，用地面积17.3hm²，
建筑面积20万m²
Planning Design of the 6#&7# Blocks of
Zhongtian New Century City, Guiyang
200　Land area: 17.3ha
Building area: 200000m²

* 半岛城邦四、五期规划设计
2007年，用地面积11.12hm²，
建筑面积40.76万m²
Planning Design of The Peninsula (Phase 4 &5)
2007　Land area: 11.12ha
Building area: 407600m²

太古城深圳蛇口东填海区城市规划设计咨询
2007年，用地面积6.6hm²，
建筑面积23.40万m²
Planning Consultation of Grand Residential
Community in the East Filling-sea Area,
Shekou, Shenzhen
2007　Land area: 6.6ha
Building area: 234000m²

深圳中信惠州东江新城一期项目规划与建筑设计
2007年，用地面积25hm²，
建筑面积60万m²
Planning & Architecture Design of CITIC
Dongjiang New City (Phase 1), Huizhou
2007　Land area: 25ha
Building area: 600000m²

* 鹏基惠州半山名苑居住区设计
2007年，用地面积49.76hm²，
建筑面积30万m²
Planning Design of Pengji Banshan Residential
Community, Huizhou
2007　Land area: 49.76ha
Building area: 300000m²

吉林省长春市科技文化综合中心概念性规划
2006年，用地面积155.8hm²，
建筑面积67.4万m²
Conceptual Planning of Science
& Culture Center, Changchun
2006　Land area: 155.8ha
Building area: 674000m²

长沙中信新城概念规划与建筑设计
2006年，用地面积109.54hm²，
建筑面积135.47万m²
Conceptual Planning & Architecture Design
of CITIC New City, Changsha
2006　Land area: 109.54ha
Building area: 1354700m²

苏州中信太湖文化论坛规划设计
2006年，用地面积55.6hm²，
建筑面积67.5万m²
Planning Design of CITIC Taihu Lake
Cultural Forum, Suzhou
2006　Land area: 55.6ha
Building area: 675000m²

深圳迈瑞研发基地周边街区概念性城市研究
2006年，用地面积2.39hm²，
建筑面积9.49万m²
Conceptual Urban Planning of
the Surrounding Area of Mindray R&D Base
2006 Land area: 2.39ha
Building area: 94900m²

深圳CEO创意领地概念规划设计
2006年，用地面积40hm²，
建筑面积33万m²
Conceptual Planning of
Creative Experiencing Origin, Shenzhen
2006 Land area: 40ha
Building area: 330000m²

深圳市蛇口东角头片区城市设计
2006年，用地面积32.70hm²，
建筑面积54.6万m²
Urban Design of DongJiaoTou Subdistrict, Shekou, Shenzhen
2006 Land area: 32.70ha
Building area: 546000m²

东莞松山湖项目概念规划设计
2006年，用地面积9.35hm²，
建筑面积12.46万m²
Conceptual Planning of
Songshanhu Residential Community, Dongguan
2006 Land area: 9.35ha
Building area: 124600m²

* 成都天府华侨城概念性规划国际竞标
2005年，国际竞标，用地面积200hm²，
建筑面积200万m²
Conceptual Planning of Tianfu-OCT, Chengdu
2005 International bidding
Land area: 200ha
Building area: 2000000m²

深圳招商华侨城尖岗山商业中心规划设计
2005年，国际竞标，用地面积6.86hm²，
建筑面积6.5万m²
Conceptual Design of
OCT Jiangganshan Shopping Center, Shenzhen
2005 International bidding
Land area: 6.86ha
Building area: 65000m²

* 成都阳明山庄规划设计
2005年，用地面积33.34hm²，
建筑面积26.56万m²
Planning Design of Chengdu Yangming
Shanzhuang Residence
2005 Land area: 33.34ha
Building area: 265600m²

北京密云休闲度假小区概念性规划设计
2005年，用地面积8hm²，
建筑面积14.28万m²
Conceptual Planning of Beijing
Miyun Vacation Residence
2005 Land area: 8ha
Building area: 142800m²

杭州市钱江科技创业中心规划设计
2005年，用地面积11.5hm²，
建筑面积11万m²
Planning Design of
Qianjiang Technology Creative Center
2005 Land area: 11.5ha
Building area: 110000m²

* 成都天府长城二期规划设计
2005年，中标方案，用地面积6.22hm²，
建筑面积21万m²
Planning Design of
Tianfu Great Wall Residence (Phase 2), Chengdu
2005 Winning project Land area: 6.22ha
Building area: 210000m2²

深圳市南澳海景度假中心概念性规划设计
2005年，用地面积64hm²，
建筑面积20万m²
Conceptual Planning Design of
Nan'ao Seascape Vacation Center
2005 Land area: 64ha
Building area: 200000m²

长沙苗圃项目概念规划设计
2005年，前期研究，用地面积80hm²
Conceptual Planning of Miaopu Project, Changsha
2005 Land area: 80ha

深圳市宝安26区旧城改造项目总体规划设计
2005年，国际竞标，用地面积23.13hm²，
建筑面积80.7万m²
General Planning for Renovation of 26# Block in Bao'an, Shenzhen
2005 International bidding,
Land area: 23.13ha
Building area: 807000m2²

深圳蛇口东"宝能-太古城"初步规划概念设计
2005年，用地面积6.93hm²，
建筑面积18.25万m²
Conceptual Planning of
Baoneng Taigu Town, Shekou, Shenzhen
2005 Land area: 6.93ha
Building area: 182573m²

* 合肥澜溪镇A、B区建筑及规划设计
2005年，用地面积7.23hm²，
建筑面积10.63万m²
竣工日期：2007年，获奖项目
Planning Design of Section A&B of Nancy Town, Hefei
2005 Land area: 7.23ha
Building area: 106300 m²
Constructed in 2007 Awarded project

长沙天际岭项目规划建筑设计
2005年，国际竞标，用地面积38.81hm²，
建筑面积39.77万m²
Planning Design of Tianjiling Project, Changsha
2005 International bidding
Land area: 38.81ha
Building area: 397706m²

深圳市金光华"绿谷蓝溪"居住小区规划设计
2005年，用地面积19.92hm²，
建筑面积35.83万m²
Planning Design of Jingguanghua
'Lvgu Lanxi' Residential Community, Shenzhen
2005 Land area: 19.92ha
Building area: 358354m²

苏州工业园区"左岸山庭"规划与建筑概念设计
2005年，国际竞标，用地面积19.53hm²，
建筑面积39.07万m²
Planning Design of Hill-Yard on Left-Bank, Suzhou,
2005 International bidding
Land area: 19.53ha Building area: 390705m²

西安中海华庭居住小区规划设计
2004年，用地面积5.41hm²，
建筑面积17.3万m²
Planning Design of Zhonghai Huating Residential
Community, Xi'an
2004 Land area: 5.41ha
Building area: 173000m²

* 成都市"中海-国际社区"规划与建筑设计
2004年，用地面积132.42hm²，
建筑面积125万m²
Planning Design of
'Zhonghai-International Community', Chengdu
2004 Land area: 132.42ha
Building area: 1250000m²

* 华润置地成都翡翠城汇锦云天居住小区规划设计
2004年，国际竞标第二名，二期用地面积8hm²，
建筑面积14.56万m²，竣工日期：2007年
Detailed Planning of CRL Emerald City
Residence (Phase 2), Chengdu
2004, Second prize of international bidding,
Land area of Phase 2: 8ha
Building area: 145600m² Constructed in 2007

* 深圳市宝安西乡富通居住区概念规划设计
2004年，国际竞标，用地面积25.23hm²，
建筑面积50万m²
Planning Design of Futong Residential Community of
Bao'an District, Shenzhen
2004 Land area: 25.23ha
Building area: 500000m²

深圳市鸿荣源龙岗中心城规划设计
2004年，国际竞标，用地面积40hm²，
建筑面积60万m²
Planning Design of Hongrongyuan Longgang
Central City, Shenzhen
2004 International bidding
Land area: 40ha
Building area: 600000m²

韶关风度广场概念性规划设计
2004年，用地面积3.2hm²，
建筑面积10.25万m²
Planning Design of Fengdu Plaza, Shaoguan
2004 Land area: 3.2ha
Building area: 102500m²

北京宣武区国信大吉片规划设计
2004年，用地面积43.9hm²，
建筑面积165万m²
Planning Design of Guoxin Dajipian in Xuanwu District,
Beijing
2004 Land area: 43.9ha
Building area: 1650000m²

湖南"天健长沙芙蓉中路项目"居住区规划设计
2004年，用地面积17.12hm²，
建筑面积23.90万m²
Planning Design of Tianjian Furongzhong
Road Residence, Changsha
2004 Land area: 17.12ha
Building area: 239000m²

福州市登云山庄总体规划设计
2004年，用地面积267.4hm²，
建筑面积68.16万m²
Planning Design of Dengyun House, Fuzhou
2004 Land area: 267.4ha
Building area: 681600m²

* 深圳市宝安区26区商业公园详细蓝图设计
2004年，用地面积26.7hm²，
建筑面积65万m²，预计竣工日期：2010年
Detailed Blueprint Design of Commercial
Park of Bao'an 26# Block, Shenzhen
2004 Land area: 26.7ha
Building area: 650000m²
Construction will be finished on 2010

中央音乐学院珠海分校规划设计
2003年，国际竞标，用地面积43hm²，
建筑面积15万m²
Planning Design of National Music University-
Zhuhai Branch
2003 International bidding
Land area: 43ha Building area: 150000m²

杭州下沙开发区城市规划设计
2003年，用地面积85hm²，
建筑面积91万m²
Urban Planning of Xiasha
Development Zone, Hangzhou
2003 Land area: 85ha
Building area: 910000m²

福建晋江市行政中心规划设计
2003年，用地面积24hm²，
建筑面积8.5万m²
Planning Design of Jinjiang
Administration Center, Fujian
2003 Land area: 24ha
Building area: 85000m²

* 苏州东城郡住宅小区规划设计
2003年，用地面积6.5hm²，
建筑面积14.70万m²，竣工日期：2006年
Planning Design of Residential Community
East County of Suzhou Industry Park
2003 Land area: 6.5ha
Building area: 147000m² Constructed in 2006

深圳盐田中心区城市规划设计
2003年，城市规划设计面积18hm²，
环境景观设计面积6.50hm²
Urban Planning of Central Area of Yantian District,
Shenzhen
2003 Urban planning area: 18ha
Landscape design area: 6.50ha

* 深圳蛇口填海区大型住宅区详细蓝图规划设计
2003年，用地面积325hm²
Detailed Blueprint Design of Large Scale Residential
Community in Shekou reclamation Area, Shenzhen
2003 Land area: 325ha

珠海情侣路滨海带规划设计
2003年，国际竞标，规划环境景观面积150hm²，
海岸线6.5公里
Planning Design of Seabelt Area of
Lovers-Road, Zhuhai
2003 International bidding
Landscape design area:150ha
Coastal line : 6.5km

深圳市沙河世纪山谷住宅小区详细蓝图规划设计
2003年，用地面积18hm²，
建筑面积67万m²
Detailed Blueprint Design of Shahe Century
Valley Residential Community, Shenzhen
2003 Land area: 18ha
Building area: 670000m²

海南省三亚阳光海岸城市规划设计
2003年，国际竞标，用地面积146.67hm²
Urban Planning of Sanya Sunny-coast, Hainan
2003 International bidding
Land area: 146.67ha

深圳市南山区大冲住宅区概念规划设计
2003年，用地面积67hm²，
建筑面积132万m²
Conceptual Planning of
Dachong Residential Community, Shenzhen
2003 Land area: 67ha
Building area: 1320000m²

苏州新加坡工业园区东湖大郡二期住宅小区规划设计
2003年，用地面积10hm²，
建筑面积18万m²
Planning Design of East Lake Residence (Phase 2) of
Singapore Industry Park, Suzhou
2003 Land area: 10ha
Building area: 180000m²

深圳市华侨城集团惠州温泉度假村规划设计
2002年，用地面积700hm²
Planning Design of OCT
Spring Spa Holiday Village, Huizhou
2002 Land area: 700ha

* 福州融侨"江南水都"首期住宅小区规划设计
2002年，用地面积12.34hm²,
建筑面积18.4万m²，竣工日期: 2006年
Detailed planning of Rongqiao
"Jiangnan Water Town" (Phase 1), Fuzhou
2002 Land area: 12.34ha
Building area: 184000m2 Constructed in 2006

福建招商局漳州开发区生活中心规划设计
2002年，规划面积100hm²,
首期商场用地面积4.4hm²，建筑面积4.80万m²
Planning Design of Living Center of Merchants Bureau
Zhangzhou Developing Zone, Fujian
2002 Planning area: 100ha
Shopping center site area of Phase 1: 4.4ha
Building area: 48000m²

南京将军山别墅及多层住宅小区规划设计
2002年，用地面积23.5hm²,
建筑面积16万m²
Planning Design of Mt. General Villa and
Muli-floors Residence, Nanjing
2002 Land area: 23.5ha
Building area: 160000m²

上海嘉定高尔夫社区规划设计
2002年，用地面积667hm²
Urban Design of
Jiading Golf Community, Shanghai
2002 Land area: 667ha

深圳高新技术开发区中心区规划设计
2001年，用地面积25.6hm²,
建筑面积27万m²
Planning Design of
Hi-tech Developing Area Center, Shenzhen
2001, Land area: 25.6ha
Building area: 270000m²

* 深圳市南海益田详细蓝图规划设计
2001年，国际竞标中标方案，用地面积31hm²,
建筑面积92万m²
Detailed Blueprint Design of
Nanhai Yitian Residence, Shenzhen
2001 Winning project of international bidding,
Land area: 31ha
Building area: 920000m²

温州苍南新市中心区规划设计
2001年，用地面积454hm²,
建筑面积15万m²
Planning Design of
Cangnan New City Center, Wenzhou
2001 Land area: 454ha
Building area: 150000m²

深圳国际网球中心俱乐部小区规划设计
2001年，用地面积9.8hm²，住宅建筑面积14万m²,
国际网球中心建筑面积3万m²
Planning Design of Shenzhen International
Tennis Center Club Residence
2001 Land area: 9.8ha
Building area of house: 140000m² Building
area of international tennis center: 30000m²

成都万安国际社区规划设计
2001年，概念设计，用地面积266hm²,
住宅建筑面积106万m²
Planning Design of Wan'an International
Community, Chengdu
2001, Land area:266ha
Housing area: 1060000m²

成都长城地产五洲花园住宅区规划设计
2001年，国际竞标，用地面积63hm²,
总建筑面积80万m²
Planning Design of Changcheng Five Continents
Residential Garden, Chengdu
2001 International bidding
Land area: 63ha
Building area: 800000m²

* 贵阳中天集团世纪新城联排住宅小区规划设计
2001年，建筑面积5.2万m²，竣工日期: 2003年
Planning Design of Zhongtian Century
New City Townhouse, Guiyang
2001 Building area: 52000m²
Constructed in 2003

吉林市松花江两岸总体规划构思及重点地区城市设计
2001年，国际竞标，规划陆地面积2525hm²,
城市设计范围971hm²
General Planning of Songhuajiang
River Bank & Urban Design of Key Areas
2001 International bidding
Land area: 25.25ha
Urban design area: 971ha

济南市南部新城区城市规划设计
2001年，用地面积396m²,
建筑面积15万m²
Urban Planning of New South District of Jinan
2001 Land area: 396ha
Building area: 150000m²

深圳罗湖百仕达山水城综合住宅小区规划设计
2001年，用地面积10hm²，住宅建筑面积28万m²,
商业建筑面积6.10万m²，酒店式公寓5万m²
Planning Design of Luohu Baishida Mountain
& Water City Residence, Shenzhen
2001 Land area:10ha Housing building area: 280000m²,
Business building area: 61000m²,
Hotel-style Apartment: 50000m²

广州新世界地产禺东西路规划设计
2001年，建筑面积14万m²
Planning Design of Yudong Xi Road Project of
New World Real Estate, Guangzhou
2001 Building area: 140000m²

深圳宝安中心区城市规划设计
2000年，国际竞标，用地面积100hm²,
建筑面积8300万m²，绿化面积30万m²
Urban Planning of Central Area of Bao'an
District, Shenzhen
2000 International bidding
Land area: 100ha Building area: 8300m²
Green area: 300000m²

东莞御花园住宅小区规划设计
2000年，建筑面积50万m²
Planning Design of Imperial
Garden Residence, Dongguan
2000 Building area: 500000m²

深圳市盐田区大梅沙梅西谷别墅区可行性研究
2000年，用地面积14hm²
Feasibility research for Meixi Valley Villa Area
of Dameisha, Shenzhen
2000 Land area: 14ha

北京万科青青家园住宅小区规划设计
2000年，用地面积23hm²,
建筑面积28万m²
Planning Design of Vanke Qingqing House, Beijing
2000 Land area: 23ha
Building area:280000m²

湖州市仁皇山区城市规划设计
2000年，国际竞标，用地面积400hm²
Urban Planning of Renhuangshan New District,
Huzhou
2000 International bidding
Land area: 400ha

南海市海八路以北规划设计
2000年，用地面积174hm²，
建筑面积23.90万m²
Planning Design of the North of haiba Road, Nanhai
2000 Land area:174ha
Building area: 239044 m²

深圳盐田碧海名峰（现名天琴湾）别墅区规划设计
1999年，用地面积29hm²，建筑面积3.8万m²，
预计竣工日期：2010年
Urban Design of
Yantian Bihaimingfeng (Lyra Bay), Shenzhen
1999 Land area: 29ha Building area: 38000 m²
Construction will be finished in 2010

广西桂林市水系规划设计
1999年，国际竞标，300公顷核心范围和
600hm²影响范围
Planning Design of
Guilin Water System, Guangxi
1999 International bidding
Core area:300ha Influenced area: 600ha

* 深圳盐田菠萝山海水温泉休闲区规划设计
1999年，用地面积24hm²，
建筑面积8万m²，竣工日期：2005年
Planning Design of Yantian Pineapple Mountain
Spa Leisure Area, Shenzhen
1999, Land area: 24ha
Building area: 80000m² Constructed in 2005

杭州市钱塘江两岸新市中心区规划设计
1999年，国际竞标，用地面积300hm²
Planning Design of New CBD along the
Qiantangjiang River, Hangzhou
1999 International bidding
Land area: 300ha

* 珠海珠澳海关地区规划设计
1998年，规划用地面积60hm²，
建筑面积7万m²，竣工日期：1998年
Planning Design of Zhu'ao Customs Area, Zhuhai
1998 Planning Land area: 60ha
Building area: 70000m² Constructed in 1998

珠海二横琴湾规划设计
1997年，用地面积100hm²
Planning Design of Second Hengqin Bay, Zhuhai
1997 Land area: 100ha

广西壮族自治区来宾法国电力公司厂区规划设计
1997年，建筑面积3万m²
Planning Design of Laibin French
Power Company Workshop, Guangxi
1997 Building area: 30000m²

江苏东海市圣戈班玻璃工厂和生活基地
法国专家村规划设计
1997年，用地面积18hm²，建筑面积2.5万m²
Planning Design of St. Gorban Glass Manufacturer
and Living Area French Expert Village, Donghai,
Jiangsu
1997 Land area: 18ha
Building area: 25000m²

江苏省无锡法国达能矿泉水厂和生活基地规划设计
1997年，建筑面积3万m²
Planning Design of French Daneng Mineral Water
Factory and Living Area, Wuxi, Jiangsu
1997 Building area: 30000m²

江苏省南京市师范大学校园区规划设计
1997年，国际竞标二等奖，
建筑面积25万m²
Planning Design of Nanjing Normal University
Campus, Jiangsu
1997 Second prize of international bidding
Building area: 250000m²

珠海中国国际青少年活动中心规划设计
1997年，用地面积98hm²
Planning Design of
China International Youth Center, Zhuhai
1997 Land area: 98ha

* 深圳市华侨城中西部城市综合区设计
1996年，用地面积85hm²，
建筑面积185万m²
Urban Planning of
OCT Center-West Area, Shenzhen
1996 Land area: 85ha
Building area: 1850000m²

深圳市福田新市中心和市民中心规划设计
1996年，国际竞标，用地面积200hm²，
市政厅建筑面积10万m²
Planning Design of Futian New City
Center and Citizen Center
1996 International bidding Land area: 200ha
Building area of City Hall: 100000m²

中山市"一河两岸"规划设计
1996年，用地面积100hm²
Planning Design of
Both Shores of One River, Zhongshan
1996 Land area: 100ha

北京国门广场住宅区规划设计
1995年，建筑面积28万m²
Planning Design of National Gate
Square Residence, Beijing
1995 Building area: 280000m²

上海市浦东六里现代化生活居住区园区规划设计
1995年，建筑面积30万m²
Planning Design of Pudong Six Li Modern
Life Residence Community, Shanghai
1995 Building area: 300000m²

珠海横琴行政中心规划设计
1994年，用地面积250hm²
Planning Design of
Hengqin Administration Center, Zhuhai
1994 Land area: 250ha

海南省海口市新城市中心规划中标方案
1994年，国际竞标中标方案，
用地面积200hm²，建筑面积350万m²
Planning Design of
New City Center of Haikou, Hainan
1994 Winning project of international bidding
Land area: 200ha Building area: 3500000m²

* 珠海大学校园规划设计
1993年，国际竞标一等奖，用地面积176hm²，
建筑面积36万m²，竣工日期：2000年
Planning Design of Zhuhai University
1993 First Prize of International bidding
Land area: 176ha
Building area: 360000m² Constructed in 2000

公建
Public building

项目总数：145
已竣工或建造中项目：37

* 已竣工或建造中项目
* Projects constructed or under construction

珠海港珠澳大桥·香港口岸国际概念设计竞赛
2010年，国际竞标，用地面积 10 hm²，
建筑面积40万m²
HongKong-Zhuhai-Macao Bridge · Hong kong Boundary Crossing Facilities International Design Ideas Competition
2010 International bidding Land area:10 ha
Building area: 400 000 m²

深圳市南山文化（美术）馆方案设计
2010年，竞标方案，用地面积 0.61 hm²，
建筑面积2.8万m²，高度40m
Architecture Design of Nanshan Culture(Arts) Museum of Shenzhen
2010 Bidding project Land area: 0.61 ha
Building area: 28 000 m² Height: 40 m

深圳创维"半导体设计中心"建筑方案设计
2010年，国际竞标第二名，用地面积 1.7 hm²，
建筑面积12.6万m²，高度100m
Architecture Design of Skyworth Semiconductor Design Center of Shenzhen
2010 Second place of international bidding
Land area: 1.7 ha Building area: 126 000 m²
Height: 100 m

深圳市高新技术企业联合总部大厦方案设计
2010 年，设计竞标第二名，用地面积 0.5 hm²，
建筑面积8.8万m²，高度100m
Architecture Design of United Headquarter of High-Tech Enterprises of Shenzhen
2010 Second place of international bidding
Land area: 0.5 ha Building area: 88 000 m²
Height: 100 m

深圳地铁一号线深大站综合体上盖物业方案设计
2010年，国际竞标，用地面积 0.98 hm²，
建筑面积9.8万m²，高度200m
Architecture Design of Building Complex above Shenzhen University Station of Metro Line No.1
2010 International bidding Land area: 0.98 ha
Building area: 98 000 m² Height: 200 m

深圳移动生产调度中心大厦概念性方案设计
2010年，国际竞标，用地面积 0.66 hm²，
建筑面积1.93万m²，高度50m
Architecture Design of Shenzhen Mobile Tower
2010 International bidding Land area: 0.66 ha
Building area: 19 300 m² Height: 50 m

深圳罗湖档案管理中心建筑设计
2010年，设计竞标第二名，用地面积 0.66 hm²，
建筑面积3.5万m²，高度60m
Architecture Design of Luohu Archives Center Shenzhen
2010 Second place of international bidding
Land area: 0.66ha Building area: 35 000 m²
Height: 60 m

深圳实验学校中学部扩建工程方案设计
2010年，竞标方案，用地面积 0.70 hm²，
建筑面积1.84万m²，高度46m
Architecture Design of Expansion Project of Shenzhen Experimental Middle School
2010 Bidding project Land area: 0.70 ha
Building area: 18 400 m² Height: 46 m

深圳创业投资大厦建筑设计
2010年，竞标方案，用地面积 0.52 hm²，
建筑面积8.06万m²，高度150m
Architecture Design of VC & PE Tower of Shenzhen
2010 Bidding project Land area: 0.52 ha
Building area: 80 600 m² Height: 150 m

深圳市青少年活动中心改扩建方案设计
2010年，国际竞标，用地面积 2.8 hm²，
建筑面积6.3万m²，高度50m
Architecture Design of Shenzhen Adolescent Activities Center
2010 International bidding Land area: 2.8 ha
Building area: 63 000 m² Height: 50 m

深圳市鼎和国际大厦方案设计
2010年，竞标方案，用地面积 0.82 hm²，
建筑面积10.6万m²，高度200m
Architecture Design of Shenzhen Dinghe International Tower
2010 Bidding project Land area: 0.82 ha
Building area: 106 000 m² Height: 200 m

深圳南山建工村保障性住房建设规划建筑设计
2010年，竞标方案，用地面积 22.42 hm²，
建筑面积47万m²，高度60m
Architecture Design of Resettlement Housing of Jiangong Village, Nanshan District, Shenzhen
2010 Bidding project Land area: 22.42 ha
Building area: 470 000 m² Height: 60 m

* 深圳南海中学建筑方案设计
2010年，中标方案，用地面积 2.4公顷，
建筑面积1.7万m²，高度24m
Architecture Design of Nanhai Middle School
2010 Winning project Land area: 2.4 ha
Building area: 17 000 m² Height: 24 m

* 河源深业东江波尔多皇家庄园规划建筑设计
2010年，国际竞标中标方案，用地面积 85 hm²，
建筑面积51.4万m²，高度100m
Conceptual Planning of Shum Yip·Bordeaux Royal Manor（Living Service Area of Dongyuan County of Heyuan, Guangdong）
2010 Winning project of international bidding
Land area: 85ha Building area: 514 000 m²
Height: 100 m

* 贵阳市城乡规划展览馆建筑设计
2010年，委托设计，用地面积3.16hm²，
建筑面积20万m²
Architectural Design of Guiyang Urban & Rural Planning Exhibition Center
2010 Mandated project Land area: 3.16 ha
Building area: 200 000 m²

* 鹏基半山名苑1期北区别墅立面施工图修改
2010年，委托设计，用地面积6hm²，
建筑面积2.3万m²
Construction Drawings Modification for the Villa Façade of Pengji Hillside Villas (Phase 1)
2010 Mandated project Land area: 6ha
Building area: 23 000m²

鹏瑞中心·深圳湾1号建筑设计
2010年，委托设计，用地面积4.6hm²，
建筑面积27万m²
Architecture Design of Pan One Center-Shenzhen Bay NO.1
2010 Mandated project Land area: 4.6ha
Building area: 270 000m²

辽宁广电中心及北方传媒文化产业园建筑方案设计
2010年，委托设计，用地面积20hm²，
建筑面积50万m²
Architectural Design of Liaoning Broadcasting Center & Northern Media Culture Park
2010   Land area: 20ha
Building area: 500 000m²

* 贵阳十里花川概念性建筑设计项目研究一
2010年，内部研究
Conceptual Design of Guiyang "Ten Miles Flower Valley"
2010 Research project

* 贵阳国际会议展览中心建筑工程设计
2009年，委托设计，用地面积51.8hm²，
建筑面积88.88万m²
Architectural Design of Guiyang International Conference & Exposition Center
2009 Mandated project Land area: 51.8 ha
Building area: 888800m²

深圳国际能源大厦建筑设计
2009年，国际竞标，用地面积0.64hm²，
建筑面积11.7万m²
Architecture Design of Shenzhen International Energy Mansion
2009 International bidding Land area: 0.64ha
Building area: 117000m²

惠州仲恺高新区总部经济区建筑设计
2009年，竞标方案，用地面积5.42hm²，
建筑面积17.55万m²
Planning & Architectural Design of Headquarters Economic Zone of Huizhou Zhongkai High-Tech Industrial Park
2009 Bidding project Land area: 5.42ha
Building area: 175500m²

深圳文学艺术中心建筑设计
2009年，竞标方案第二名，用地面积1.0hm²，
建筑面积6.20万m²，高度100m
Architectural Design of Shenzhen Literary Arts Center
2009 Second prize of bidding
Land area: 1.0ha
Building area: 62000m² Height: 100m

深圳移动生产调度中心大厦概念性方案设计
2009年，竞标方案，用地面积0.56hm²，
建筑面积10万m²，高度159.6m
Architectural Design of Production Scheduling Center Building of China Mobile (Shenzhen)
2009 Bidding project Land area: 0.56ha
Building area: 100000m² Height:159.6m

深圳远致创业园建筑设计
2009年，竞标方案，用地面积12.17hm²，
建筑面积60.67万m²，高度245m
Planning & Architectural Design of Yuanzhi Innovation Park, Shenzhen
2009 Bidding project Land area: 12.17ha
Building area: 606700m² Height: 245m

深圳高新园区软件产业基地建筑设计（第三、四标段）
2009年，竞标方案，用地面积14.9hm²，
建筑面积61万m²
Planning & Architectrual Design of Software Industry Base (section III & IV) of Shenzhen High-tech Industrial Park
2009 Bidding project Land area: 14.9ha
Building area: 610000m²

深圳中信银行大厦建筑设计
2009年，竞标方案，用地面积0.44hm²，
建筑面积6.58万m²
Architectrual Design of China CITIC BANK (shenzhen) Mansion
2009 Bidding project Land area: 0.44ha
Building area: 65800m²

南方科技大学和深圳大学新校区拆迁安置项目
—产业园区建筑设计
2009年，竞标方案，用地面积15.25hm²，建筑面积63.10万m²
Removing and Resettlement of New Campus of South University of Science and Technology and Shenzhen University -Planning & Architectural Design of Industry Park
2009 Bidding project Land area: 15.25ha
Building area: 631000m²

* 深圳地铁竹子林车辆段改扩建工程上盖建筑设计
2009年，中标方案，用地面积1.90hm²，
建筑面积4.42万m²
Architectural Design for Buildings above Zhuzilin Metro Station of Shenzhen
2009 Winning project Land area: 1.90ha
Building area: 44200m²

深圳中广核大厦建筑设计
2008年，国际竞标，用地面积1.01hm²，
建筑面积18.85万m²，高度159.98m
Architecture Design of China Guangdong Nuclear Power Industry's Office Tower in Shenzhen
2008 International bidding Land area :1.01ha
Building area: 188500 m² Height:159.98m

深圳虚拟大学园院校产业化综合大楼建筑设计
2008年，用地面积0.78hm²，
建筑面积3.2万m²
Architecture Design of Complex Office of Shenzhen Virtual University Park
2008 Land area: 0.78ha
Building area: 32000m²

深圳人才园建筑设计
2008年，用地面积3.61hm²，
建筑面积8.32万m²
Architecture Design of Shenzhen Human Resouces Park
2008 Land area:3.61ha
Building area: 83200m²

珠海歌剧院建筑设计
2008年，国际竞标，用地面积42hm²，
建筑面积4.3万m²
Architecture Design of Zhuhai Opera House
2008 International bidding
Land area: 42ha
Building area: 43000m²

深圳航天国际中心建筑设计
2008年，国际竞标，用地面积1.05hm²，
建筑面积15万m²，高度241.2m
Architecture Design of International Aerospace Center of Shenzhen
2008 International bidding Land area: 1.05ha
Building area: 150000m² Height: 241.2m

成都怡湖玫瑰湾建筑设计
2008年，用地面积12.57hm²，
建筑面积62.5万m²
Architectual Design of Yihu Rose Bay Community, Chengdu
2008 Land area: 12.57ha
Building area: 625000m²

深圳市光明新区公明文化艺术和体育中心建筑设计
2008年，用地面积9.9hm²，
建筑面积5.09万m²
Architecture Design of
Gongming Culture, Art & Sports Center of
Guangming New Town, Shenzhen
2008 Land are: 9.9 ha
Building area: 50900m²

深圳南澳鹅公湾水产养殖基地3#地建筑设计
2008年，用地面积1.61hm²，
建筑面积0.64万m²
Architecture Design of
Aquiculture Base Plot 3 of Egong Bay in
Nan'ao, Shenzhen
2008 Land area: 1.61ha
Building area: 6400m²

深圳宝安中心区N28区九年一贯制学校建筑设计
2008年，用地面积2.7hm²，
建筑面积2.42万m²
Architecture Design of
Nine Year's Volunteer Educational school of
Bao'an Block N28, Shenzhen
2008 Land area: 2.7ha
Building area: 24200m²

深圳市龙岗区南湾中学建筑设计
2008年，用地面积2.75hm²，
建筑面积1.66万m²
Architecture Design of
Nine Year's Volunteer Educational school
Longgang Nanwan in Shenzhen
2008, Land area: 2.75ha
Building area: 16600m²

宿州综合体概念性建筑设计
2008年，用地面积21.3hm²，
建筑面积67.92万m²
Architecture Design of Suzhou Complex
2008 Land area:21.3ha
Building area: 679200m²

* 顺德深业城一期建筑立面设计
2008年，建筑面积18万m²
Architecture Facade Design of
Shum Yip's Community (Phase 1), Shunde
2008 Building area: 180000m²

中国饮食文化城建筑设计
2008年，用地面积220hm²，
建筑面积38万m²
Architecture Design of
Chinese Gastrologic and Cultural Town of Shenzhen
2008 Land area: 220ha
Building area: 380000m²

深圳市南油购物公园建筑设计
2008年，国际竞标第一名
用地面积13hm²，建筑面积45.4万m²
Architecture Design of
Nanyou Shopping Park, Shenzhen
2008, First prize of international bidding
Land area: 13ha Building area: 454000

深圳光明新城万丈坡居住小区建筑设计
2008年，详蓝编制范围22hm²，
规划研究范围49.6hm²，建筑面积53.23万m²
Architecture Design of Wanzhangpo Residential
Community of Guangmin New Town, Shenzhen
2008 Planning area: 22ha
Research area: 49.6ha Building area: 532300m²

珠海五洲花城二期建筑设计
2008年用地面积16.5hm²，
建筑面积60万m²
Architecture Design of
Five Continental Residence Garden, Zhuhai
2008 Land area: 16.5ha
Building area: 600000m²

* 深圳康佳研发大厦建筑设计 方案一
2007年，中标方案，用地面积0.96hm²，
建筑面积7.84万m²，高度100m
Architecture Design of
Konka R&D Building, Shenzhen (Schema I)
2007 Winning project Land area: 0.96ha
Building area: 78400m² Height: 100m

深圳康佳研发大厦建筑设计 方案二
2007年，用地面积0.96hm²，
建筑面积7.93万m²，高度100m
Architecture Design of
Konka R&D Building , Shenzhen (Schema
2007 Land area: 0.96ha
Building area: 79300m² Height: 100m

* 深圳市长富金茂大厦建筑设计
2007年，中标方案，用地面积1.88hm²，
建筑面积22.19万m²，高度350m
Architecture Design of
Changfu Jinmao Building, Shenzhen
2007 Winning project Land area: 1.88ha,
Building area: 221900m² Height: 350m

* 深圳龙岗区天安数码新城建筑设计
2007年，中标方案，用地面积4.63hm²，
建筑面积24.68万m²
Architecture Design of
Longgang District Cyber Park, Shenzhen
2007 Winning project Land area: 4.63ha
Building area: 246800m²

狮山客运站建筑设计
2007年，中标方案，用地面积9.5hm²，
建筑面积11.45万m²
Architecture Design of
Shishan Long Distance Bus Station
2007 Winning project Land area: 9.50ha,
Building area: 114500m²

深业集团惠州开发总部建筑方案设计
2007年，用地面积0.9hm²，
建筑面积7.93万m²
Architecture Design of Huizhou Developme
Headquarters of Shum Yip Group
2007 Land area: 0.9ha
Building area: 79300m²

中国太湖文化论坛会议中心建筑方案设计
2007年，用地面积2.5hm²，
建筑面积2.94万m²
Architecture Design of
Convention Center of Taihu Lake Cultural Forum
2007 Land area: 2.5ha
Building area: 29400m²

* 深圳市贝丽中学建筑方案设计
2007年，用地面积1.95hm²，
建筑面积2.53万m²
Architecture Design of
Shenzhen Beili School
2007 Land area: 1.95ha
Building area: 25300m²

* 贵阳中天世纪新城中心商业建筑设计
2007年，用地面积5.05hm²，
建筑面积9.7万m²
Architecture Design of the Central Commerce of
Guiyang Zhongtian New Century City
2007 Land area: 5.05ha
Building area: 97000m²

* 成都华润置地翡翠城小学建筑方案设计
2007年，用地面积1.65hm²，
建筑面积1.71万m²
Architecture Design of
CRL Jade-City Primary School, Chengdu
2007 Land area: 1.65ha
Building area:17100m²

* 深圳市半岛城邦住宅区二期售楼中心
建筑概念方案设计
2007年，用地面积2000m²，
建筑面积1500m²
Architecture Design of Sales Center of
The Peninsula (Phase 2), Shenzhen
2007 Land area: 2000m²
Building area: 1500m²

* 深圳市半岛城邦二期配套小学
建筑概念方案设计
2007年，用地面积1.04hm²，
建筑面积1.02万m²
Architecture Design of
Primary School in The Peninsula (Phase 2), Shenzhen
2007 Land area: 1.04ha
Building area: 10200m²

* 成都华润置地二十四城小学建筑方案设计
2007年，用地面积1.94hm²，
建筑面积1.53万m²
Architecture Design of
Primary School of CRL 24th City, Chengdu
2007 Land area: 1.94ha
Building area: 15300m²

* 万科宁波金色水岸会所建筑设计
2007年，用地面积2000m²，
建筑面积4500m²，竣工时间2008年
Architecture Design of Club-House of
Vanke Golden Waterfront Community, Ningbo
2007 Land area: 2000m²
Building area: 4500m² Constructed in 2008

佛山南海区狮山镇文化体育公园建筑设计
2007年，竞标方案，用地面积13.9hm²，
建筑面积28万m²
Architecture Design of Shishan Cultural &
Sport Park, Nanhai District, Foshan
2007 Bidding scheme Land area: 13.9ha
Building area: 280000m²

* 深圳市城建集团观澜居住区规划建筑设计
2007年，中标方案，用地面积16.99hm²，
建筑面积34.6万m²
Architecture Design of
Guanlan Residential Community, Shenzhen
2007 Winning project Land area: 16.99ha
Building area: 346000m²

* 华润置地成都翡翠城四期公共建筑设计
2007年，用地面积7.56hm²，
建筑面积41.56万m²
Architecture Design of Public Buildings of
CRL Jade-City (Phase 4), Chengdu
2007 Land area: 7.56ha
Building area: 415600m²

* 深圳市南山商业文化中心区超高层建筑设计
2006年，国际竞标中标方案，用地面积2.57hm²，
建筑面积16.3万m²，高度288m
Architecture Design of the Tower in Nanshan
Commercial & Cultural Center, Shenzhen
2006 Winning project of international bidding
Land area: 2.57ha
Building area: 163000m² Height: 288m

深圳市东海中心建筑设计
2006年，用地面积3.3hm²，
建筑面积40.3万m²，建筑高度230m
Architecture Design of
Shenzhen Donghai Center
2006 Land area: 3.3ha
Building area: 403000m² Height: 230m

南宁市城建档案馆（新馆）建筑设计
2006年，用地面积1.21hm²，
建筑面积5.48万m²
Architecture Design of
Nanning Urban Construction Archives
2006 Land area: 1.21ha
Building area: 54800m²

深圳鹏基龙电工业城建筑设计
2006年，用地面积3.08hm²，
建筑面积8.64万m²
Architecture Design of
Pengji Longdian Office Block, Shenzhen
2006 Land area: 3.08ha
Building area: 86400m²

* 深圳市沙河世纪假日广场建筑设计
2006年，用地面积1.13hm²，
建筑面积10.24万m²，竣工日期：2009年
Architecture Design of
Shahe Century Holiday Plaza, Shenzhen
2006 Land area: 1.13ha,
Building area: 102400m² Constructed in 2009

四川文化城建筑设计
2006年，用地面积1.05hm²，
建筑面积11.4万m²
Architecture Design of Sichuan Cultural City
2006, Land area: 1.05ha,
Building area: 114000m²

无锡新世界国际纺织服装市场中心商务区概念性设计
2006年，用地面积8.78hm²，
建筑面积27.6万m²
Architecture Design of Central Commercial Area of
Wuxi New World International Textile and
Clothing Market
2006 Land area: 8.78ha
Building area: 276000m²

三洋厂房改造项目概念性设计方案
2006年，用地面积1.03hm²，
建筑面积1.65万m²
Architecture Design of
SANYO Workshop Rebuild Project
2006 Land area: 1.03ha
Building area: 16500m²

* 深圳华侨城中学（高中部）建筑设计
2005年，用地面积6.3hm²，
建筑面积3.32万m²，竣工日期：2008年
Architecture Design of
Shenzhen Bay High School
2005 land area: 6.3ha
Building area: 33200m² Constructed in 2008

杭州市钱江创业中心建筑设计
2005年，用地面积4.52hm²，
建筑面积11万m²
Architecture Design of
Qianjiang Creation Center, Hangzhou
2005, Land area: 4.52ha,
Building area: 110000m²

杭州钱江新城中央商务区办公建筑概念设计
2005年，用地面积2.71hm²，
建筑面积12.45万m²
Architecture Design of
Qianjiang New CBD Office, Hangzhou
2005, Land area: 2.71ha,
Building area: 124500m²

杭州天际大厦建筑设计
2005年，国际竞标，用地面积1.13hm²，
建筑面积5.25万m²
Architecture Design of
Tianji Mansion, Hangzhou
2005 International bidding Land area: 1.13ha
Building area: 52500m²

深圳市招商华侨城尖岗山商业中心建筑设计
2005年，国际竞标，用地面积6.86hm²，
建筑面积6.53万m²
Architecture Design of
OCT Jiangganshan Shopping Center, Shenzhen
2005 International bidding Land area: 6.86ha
Building area: 65300m²

* 上海浦南中学建筑设计
2005年，用地面积1.7hm²，
建筑面积1.06万m²
Architecture Design of
Punan High School, Shanghai
2005 Land area: 1.7ha
Building area: 10600m²

无锡博物馆、科技馆、革命陈列馆综合体建筑设计
2005年，国际竞标，
用地面积2.02hm²，建筑面积4.30万m²
Architecture Design of Wuxi Complex Museum
2005 International bidding
Land area: 2.02ha Building area: 43000m²

海与建筑-餐饮中心建筑设计意向
2005年，国际竞标，用地面积0.10hm²，
建筑面积570m²
Architecture Design of Sea & Building Restaurant
2005 International bidding Land area: 0.1ha,
Building area: 570m²

韶关风度广场建筑设计
2004年，用地面积3.2hm²，
建筑面积10.25万m²
Architecture Design of
Shaoguan Feng-Du Plaza
2004 Land area: 3.2ha
Building area: 102500m²

* 深圳市南山商业文化中心区11-01地块项目概念设计
2004年，用地面积2.57hm²，
建筑面积17.53万m²
Architecture Design of
No.11-01 Plot of Nanshan Business &
Culture Center, Shenzhen
2004 Land area: 2.57ha
Building area: 175300m²

* 成都置信天府花城建筑设计
2004年，用地面积3.03hm²，
建筑面积1.4万m²，竣工日期：2007年
Architecture Design of
Zhixin Tianfu Flower City Residence
2004 Land area: 3.03ha
Building area: 14000m² Constructed in 2007

成都置信未来广场A期建筑设计
2004年，用地面积6.39hm²，
建筑面积10.87万m²，竣工日期：2005年
Architecture Design of
Zhixin "Future Plaza" (Phase A), Chengdu
2004 Land area: 6.39ha
Building area: 108700m² Constructed in 2005

广东省南海盐步信息发展中心建筑设计
2004年，用地面积6hm²，
建筑面积3万m²
Architecture Design of Yanbu
Information Development Center, Nanhai
2004 Land area: 6ha
Building area: 30000m²

深圳市宝安城区26区商业公园建筑设计
2004年，规划设计面积26.7hm²，
总建筑面积65万m²，
Architecture Design of
Commercial Park of Bao'an 26# Block, Shenzhen
2004 Planning area: 26.7ha
Building area: 650000m²

成都川投置信广场规划与建筑设计
2004年，用地面积0.84hm²，
建筑面积5.89万m²
Architecture Design of
Chuantou Zhixin Plaza, Chengdu
2004 Land area: 0.84ha
Building area: 58900m²

深圳市福田科技广场规划与建筑设计
2004年，用地面积3.87hm²，
建筑面积21.7万m²
Architecture Design of
Futian Science & Technology Plaza, Shenzhen
2004 Land area: 3.87ha
Building area: 217000m²

* 深圳欧贝特卡系统科技有限公司新工厂改造项
2004年，总改造面积3975m²
Reconstruction Project of
New Workshop of Shenzhen Obetaca
System Science & Technology Co., Ltd
2004 Total reconstruction area: 3795m²

北京宣武区国信大吉片建筑设计
2004年，用地面积43.9hm²，
建筑面积165万m²
Architecture Design of Guoxin Dajipian of
Xuanwu District, Beijing
2004 Land area: 43.90ha
Building area: 1650000m²

湖南天健长沙芙蓉中路项目居住区建筑设计
2004年，用地面积17.12hm²，
建筑面积75.12万m²
Architecture Design of Tijian Residence on
Furongzhong Road, Changsha
2006 Land area: 17.12ha
Building area: 751200m²

* 深圳鹏基集团商务时空建筑设计
2003年，用地面积1.1hm²，
建筑面积2.7万m²，竣工日期：2007年
Architecture Design of
Pengji Business Space-time Mansion, Shenzhen
2003 Land area: 1.1ha,
Building area: 27000m² Constructed in 2007

海南省三亚阳光海岸城市建筑设计
2003年，国际竞标，
总设计面积146.67hm²
Architecture Design of
Sanya Sunny-Coast City, Hainan
2003 International bidding
Total design area: 146.67ha

* 深圳龙岗下沙金沙滩滨海休闲带建筑设计
2003年，用地面积17hm²，
竣工日期：2004年
Architecture Design of
Xiasha Golden Bench Seaside Resorts
2003 Land area: 17ha
Constructed in 2004

福建晋江市行政中心建筑设计
2003年，用地面积24hm²，
建筑面积8.5万m²
Architecture Design of
Jinjiang Administration Center, Fujian
2003 Land area: 24ha
Building area: 85000m²

郑州郑东新区颐和医院建筑设计
2003年，用地面积18.07hm²，
建筑面积20.62万m²
Architecture Design of Yihe Hospital, Zhengzhou
2003 Land area: 18.07ha
Building area: 206200m²

郑州郑东新区房地产展销大楼建筑设计
2003年，国际竞标第一名，
建筑面积6.5万m²
Architecture Design of Housing Display and
Selling Building of New East District, Zhengzhou
2003 First prize of international biddin

深圳星河万豪五星级酒店建筑设计
2003年，用地面积0.90hm²，
建筑面积11万m²
Architecture Design of
Xinghe Marriot International Hotel, Shenzhen
2003 Land area: 0.90ha
Building area: 110000m²

中央音乐学院珠海分校建筑设计
2003年，用地面积43hm²，
建筑面积15万m²
Architecture Design of
National Music University- Zhuhai Branch
2003 Land area: 43ha
Building area: 150000m²

深圳市龙岗区规划展览综合大楼建筑设计
2003年，国际竞标中标方案，
用地面积4.88hm²，建筑面积3.14万m²
Architecture Design of
Longgang Planning Exhibition Complex Building, Shenzhen
2003 Winning project of international bidding
Land area: 4.88ha Building area: 31400m²

* 深圳大剧院外观改造建筑设计
2003年，国际竞标中标方案，
总建筑面积8.52万m²，竣工日期：2006年
Architectural Facade Design of
Shenzhen Grand Theatre
2003 Winning project of international bidding
Building area: 85200m² Constructed in 2006

上海法德国际学校校园建筑设计
2003年，用地面积5.7hm²，
建筑面积2.50万m²
Architecture Design of
Shanghai French-German International School
2003 Land area: 5.7ha
Building area: 25000m²

杭州广厦-邮政大厦建筑设计
2003年，用地面积1.96hm²，
建筑面积9万m²
Architectue Design of Hangzhou Post Tower
2003 Land area: 1.96ha
Building area: 90000m²

* 北京金世纪酒店建筑立面设计
2003年，建筑面积3.50万m²，
竣工日期：2006年
Architectural facade Design of
Beijing Golden Century Hotel
2003 Building area: 35000m²
Constructed in 2006

杭州浙江宾馆四星级酒店建筑设计
2003年，用地面积9hm²，
建筑面积7万m²
Architecture Design of
Zhejiang Hotel (Four-star), Hangzhou
2003 Land area: 9 ha
Building area: 70000m²

浙江电子信息大厦建筑设计
2003年，国际竞标中标方案，
用地面积0.88hm²，建筑面积7万m²
Architecture Design of
Information Mansion, Zhejiang
2003 Winning project of international bidding
Land area: 0.88ha Building area: 70000m²

杭州下沙开发区城市建筑设计
2003年，用地面积85hm²，
建筑面积91万m²
Architecture Design of
Xiasha District, Hangzhou
2003 Land area: 85 ha
Building area: 910000m²

深圳葵涌海滨度假中心建筑设计
2002年，国际竞标，
建筑面积1.5万m²
Architecture Design of
Kuiyong Seaside Holiday Center, Shenzhen
2002 International bidding
Building area: 15000m²

* 广州法国文化协会（Alliance French）室内设计
2002年，设计面积215m²，
竣工日期：2002年
Interior Design of
Guangdong Alliance French
2002 Design area: 215m²
Constructed in 2002

* 深圳半岛城邦建筑设计
2002年，用地面积5hm²，
建筑面积18万m²，竣工日期：2007年
Architecture Design of The Peninsula, Shenzhen
2002 Land area: 5 ha Building area: 160000m²
Constructed in 2007

* 成都新城市广场建筑设计
2002年，建筑面积17万m²，
竣工日期：2005年
Architecture Design of Chengdu New City Plaza
2002 Building area: 170000m²
Constructed in 2005

* 杭州南北商务港建筑设计
2002年，建筑面积3万m²，
获奖项目，竣工日期：2005年
Architecture Design of
Nanbei Business Port, HangZhou
2002 Building area: 30000m²
Award project Constructed in 2005

成都国际商城建筑设计
2002年，国际竞标中标方案，
建筑面积18万m²，竣工日期：2006年
Architecture Design of
Chengdu International Shopping Mall
2002 Winning project of international bidding
Building area: 180000m² Constructed in 2006

* 广州新世界地产凯旋新世界建筑设计
2001年，用地面积10hm²，竣工日期：2004年
Architecture Design of Guangzhou Triumph
New World Residential Community
2001 Land area:10 ha Constructed in 2004

深圳市党委建筑设计
2001年，建筑面积8万m²
Architecture Design of Shenzhen
Municipal Party Committee and Party School
2001 Building area: 80000m²

深圳国际网球中心俱乐部小区建筑设计
2001年，用地面积9.8hm²，住宅建筑面积14万m²，
国际网球中心建筑面积3万m²
Architecture Design of Shenzhen International
Tennis Center Club Residence
2001 Land area: 9.8ha
Building area of house: 140000m²
Building area of international tennis center: 30000m²

深圳高新技术开发区中心区建筑设计
2001年，用地面积25.6hm²，
建筑面积27万m²
Architecture Design of Central Area of
Shenzhen Hi-tech Development Zone
2001 Land area: 25.6ha
Building area: 270000m²

深圳国际市长大厦建筑设计
2001年，建筑面积3.20万m²
Architectue Design of
Shenzhen International Mayors Building
2001 Building area: 32000m²

深圳华侨城中央会所建筑设计
2001年，建筑面积7000m²
Architecture Design of OCT
Central Club House, Shenzhen
2001 Building area: 7000m²

* 深圳华侨城中央科教所附属9年制学校建筑设计
2001年，国际竞标中标方案，用地面积3.30hm²，
建筑面积2.30万m²，竣工日期：2003
Architecture Design of Nanshan School Affiliated with
China National Institute for Educational Research
2001 Winning project of international bidding,
Land area: 3.3ha Building area: 23000m²
Constructed in 2003

* 深圳市梧桐山电视发射塔建筑设计
2001年，国际竞标中标方案，
建筑面积1万m²，高度300m
Architecture Design of
Wutong Mountain TV Transmission Tower, Shenzhen 2001 Winning project of international bidding

温州苍南新市中心区政务中心建筑设计
2001年，规划面积454hm²，
建筑面积12万m²
Architecture Design of Administration
Center of New Cangnan CBD, Wenzhou
2001 Planning area: 454ha
Building area: 120000m²

杭州市钱塘江两岸新市中心区建筑设计
2001年，国际竞标，用地面积300m²
Architecture Design of New CBD along the
Qiantangjiang River, Hangzhou
2001 International bidding Land area: 300 ha

济南市南部新城区政务中心建筑设计
2001年，规划面积396hm²，
建筑面积15万m²
Architecture Design of Administration
Center of New South District, Jinan
2001 Planning area: 396 ha,
Building area: 150000m²

深圳创智集团办公大楼建筑设计
2001年，建筑面积3.2万m²
Architecture Design of
Chuangzhi Group Office Building, Shenzhen
2001 Building area: 32000m²

南海市海八路以北建筑设计
2001年，用地面积174hm²，
建筑面积23.90万m²
Architecture Design of
the North of Haiba Road, Nanhai
2001 Land area: 174 ha
Building area: 239000m²

吉林市松花江沿岸总体规划构思及重点地区城市建筑设计
2001年，国际竞标，规划陆地面积2525hm²，
城市设计范围971hm²
Architecture Design of the Key Area of
Songhua River Shores, Jilin
2001 International bidding
Land area: 2525ha Urban design area: 971ha

杭州西湖文化广场建筑设计
2001年，建筑面积20万m²
Architecture Design of
West Lake Cultural Plaza, Hangzhou
2001 Building area: 200000m²

* 杭州嘉华国际中心办公楼建筑设计
2001年，建筑面积5.6万m²，
获全国人居金典奖、商务类金奖，竣工日期：2003
Architecture Design of
Jiahua International Office Building, Hangzhou
2001 Building area: 56000m²
Awarded as the Golden Commercial Housing in
China Constructed in 2003

法国Decathlon公司深圳基地项目建筑设计
2000年，可行性研究，用地面积3hm²
Architecture Design of
Decathlon (France) Shenzhen Office
2000 Feasibility study
Land area: 3 ha

深圳福田新中心区香港凤凰卫视亚洲总部建筑设计
2000年，可行性研究，建筑面积8万m²
Architecture Design of
HongKong Phoenix TV's Aisa Headquarter
in Futian CBD, Shenzhen
2000 Feasibility study
Building area: 80000m²

* 安徽合肥国际会议展览中心建筑设计
2000年，国际竞标中标方案，
建筑面积5.6万m²，竣工日期：2002年
Architecture Design of Hefei International
Conference & Exhibition Center, Anhui
2000 Winning project of international bidding
Building area: 56000m² Constructed in 2002

深圳宝安中心城建筑设计
2000年，国际竞标，用地面积100hm²，
建筑面积8.3万m²，绿化面积30万m²
Architecture Design of Central District of Bao'an, Shenzhen
2000 International bidding Land area: 100ha
Building area: 83000m² Green area: 300000m²

深圳盐田沙头角体育中心建筑设计
2000年，建筑面积8000m²，
竣工日期：2004年
* Architecture Design of Shatoujiao
Sports Center of Yantian District, Shenzhen
2000 Building area: 8000m²
Constructed in 2004

深圳星彦大厦办公楼建筑设计
2000年，建筑面积4.5万m²
Architecture Design of
Xingyan Office Building, Shenzhen
2000 Building area: 45000m²

湖州市仁皇山新区城市建筑设计
2000年，国际竞标，用地面积400hm²
Architecture Design of
Renhuangshan New District, Huzhou
2000 International bidding
Land area: 400ha

* 深圳规划国土局盐田分局办公楼建筑设计
1999年，国际竞标中标方案，用地面积0.6hm²
建筑面积9469m²，竣工日期：2001年
Architecture Design of Yantian Branch of
Building of Shenzhen Urban Planning & Land Bureau
1999 Winning project of international bidding
Land area: 0.6 ha Building area: 9469m²
Constructed in 2001

* 深圳规划国土局盐田分局办公楼室内设计
1999年，设计面积7000m²，
竣工日期：2001年
Interior Design of Yantian Branch Office Building
of Shenzhen Urban Planning & Land Bureau
1999 Design area: 7000m²
Constructed in 2001

深圳地王大厦第68层凌霄阁室内设计
1999年，设计面积2000m²，
竣工日期：2003年
Interior Design of Lingxiaoge on the 68th floor of
Diwang Commercial Center
1999 Design area: 2000m²
2000 Constructed in 2003

杭州市大剧院建筑设计
1999年，国际竞标，
建筑面积4.5万m²
Architecture Design of Hangzhou Grand Theatre
1999 International bidding
Building area: 45000m²

杭州市吴山商城建筑设计
1999年，国际竞标，用地面积6hm²，
建筑面积9万m²
Architecture Facade Design of
Wushan Shopping Mall, Hangzhou
1999 International bidding
Land area: 6 ha Building area: 90000m²

深圳罗湖华佳广场建筑立面设计
1998年，竣工日期：1998年
Architecture Facade Design of
Shenzhen Luohu Huajia Plaza
1998 Constructed in 1998

* 珠海珠澳海关地区规划与口岸联检大楼建筑设计
1998年，规划用地面积60hm²，
建筑面积7万m²，竣工日期：1998年
Planning Design of Zhuhai Zhu'ao Custom Area and
Architecture Design of Port Checking Building
1998 Land area of planning: 60ha,
Building area: 70000m² Constructed in 1998

* 深圳华侨城OCT生态广场建筑设计
1998年，城市公园景观设计用地面积5hm²
建筑面积3万m²，竣工日期：1999年
Architecture Design of OCT Ecology Plaza ,Shenzhen
1998 Land area of landscape design: 5 ha
Building area: 30000m² Constructed in 1999

深圳市文化广场（现更名为中信广场）建筑设计
1998年，国际竞标，建筑面积15万m²
Architecture Design of Shenzhen Cultural Plaza
(Zhongxin City Plaza now)
1998 International bidding
Building area: 150000m²

江苏省南京市师范大学校园区建筑设计
1997年，国际竞标二等奖，
建筑面积25万m²
Architecture Design of Nanjing
Normal University, Jiangsu
1997 Second prize of international bidding,
Building area: 250000m²

北京市乐邦集团和中国电影合作制片公司
综合大厦建筑设计
1997年，建筑面积5万m²
Architecture Design of Integrated Building of
Beijing Lebang Group and
China Film Co-product Corporation
1997 Building area: 50000m²

深圳电视中心工程建筑设计
1997年，国际竞标中标方案，
建筑面积4.5万m²
Architecture Design of Shenzhen TV Tower
1997 Winning project of international bidding
Building area: 45000m²

* 深圳市华侨城中西部城市综合区城市建筑设计
1996年，城市和环境设计用地面积85hm²，
建筑面积185万m²
Architecture Design of
Shenzhen OCT Central-West Area
1996 Land area of urban and
environmental design: 85ha
Building area: 1850000m²

深圳市福田新市中心和市民中心建筑设计
1996年，国际竞标，用地面积200hm²
市政厅建筑面积10万m²
Architecture Design of Futian New City Center and
Citizen Center of Shenzhen
1996 International bidding
Land area: 200ha Building area of city hall: 100000m²

上海东方医院建筑设计
1995年，国际竞标二等奖，
建筑面积4.5万m²
Architecture Design of Shanghai Oriental Hospital
1995 Second prize of international bidding
Building area: 45000m²

浙江省杭州国际金融中心建筑设计
1994年，国际竞标二等奖，
建筑面积5万m²
Architecture Design of Hangzhou
International Finance Center, Zhejiang
1994 Second prize of international bidding
Building area: 50000m²

海南省海口市新城市中心建筑设计
1994年，国际竞标中标方案，
用地面积200hm²，建筑面积350万m²
Architecture Design of New City Center of
Haikou, Hainan
1994 Winning project of international bidding
Land area: 200ha Building area: 3500000m²

珠海横琴行政中心建筑设计
1994年，国际竞标，用地面积100hm²
Architecture Design of
Zhuhai Hengqin Administration Center
1994 Land area: 100ha

北京市长安街皇城内珠宝商业中心设计
1994年，国际竞标一等奖，
建筑面积4.5万m²
Architecture Design of Jewelry Industry Center in
Imperial City of Chang'an Street, Beijing
1994 First prize of international bidding
Building area: 45000m²

上海东方音乐厅建筑设计
1994年，国际竞标二等奖，
建筑面积4万m²
Architecture Design of Shanghai Oriental Odeum
1994 Second prize of international bidding
Building area: 40000m²

* 珠海大学校园建筑设计
1993年，国际竞标一等奖，用地面积176hm²，
建筑面积36万m²，竣工日期：2000年
Architecture Design of Zhuhai University
1993 First prize of international bidding
Land area: 176ha Building area: 360000m²
Constructed in 2000

住宅
Residence

项目总数: 73
已竣工或建造中项目: 20

* 已竣工或建造中项目
* Projects constructed or under construction

* 成都天府华侨城纯水岸C组团高层区及公共配套项目建筑设计
2009年,委托设计,用地面积2.51hm², 建筑面积11.23万m²
Architectural Design of Residence Towers &
Public Supporting Facilities of Block C of Chengdu
OCT-The Riviera
2009 Mandated project Land area: 2.51ha
Building area: 112300m²

* 深圳城建集团观澜项目建筑设计
2009年,中标方案,用地面积19.46hm²,
建筑面积55.53万m²
Architectural Design of SZEG Guanlan
Residential Community in Shenzhen
2009 Mandated project Land area: 19.46ha
Building area: 555300m²

* 中山金源花园住宅建筑设计
2009年,中标方案,用地面积21.22hm²,
建筑面积79.26万m²
Architectural Design of Jin Yuan Garden
Residence in Zhongshan
2009 Winning project Land area: 21.22
Building area: 792600m²

南方科技大学和深圳大学新校区拆迁安置项目-
商住综合区住宅建筑设计
2009年，国际竞标招标方案，用地面积17.77hm²，建筑面积60万m²
Removing and Resettlement of New Campus of South University of Science and Technology and Shenzhen University -
Architectural Design for Commerce and Residence
2009 Bidding project Land area: 17.77ha
Building area: 600000m²

* 深圳半岛城邦3期超高层住宅建筑设计
2009年，委托设计，用地面积5.68hm²，建筑面积19.07万m²
Architectural Design of Mega High Rise
Residential Tower of Shenzhen Peninsula (phase III)
2009 Mandated project Land area: 5.68ha
Building area: 190700m²

长春市净月区梧桐街住宅建筑设计
2008年，用地面积34.8hm²，建筑面积45万m²
Architecture Design of Residential Community of
Phoenix Tree Street of Jingyue District, Changchun
2008 Land area: 34.8ha
Building area: 450000m²

深圳市光明新区保障性住房工程住宅建筑设计
2008年，用地面积4.12hm²，建筑面积13万m²
Architecture Design of Social Community of
Guangming New Town, Shenzhen
2008 Land area: 4.12ha
Building area: 130000m²

成都怡湖玫瑰湾住宅建筑设计
2008年，用地面积12.57hm²，建筑面积62.5万m²
Architecture Design of Community of
Yihu Rose Bay, Chengdu
2008 Land area: 12.57ha
Building area: 625000m²

长春市高新区C-6地块住宅建筑设计
2008年，用地面积8.57hm²，建筑面积15.97万m²
Architecture Design of Residential Community of
High-Tech C-6 Block, Changchun
2008 Land area: 8.57ha
Building area: 159700m²

宿州综合体概念性建筑设计
2008年，用地面积21.3hm²，建筑面积67.92万m²
Architecture Design of Suzhou Complex
2008 Land area: 21.3ha
Building area: 679200m²

* 顺德深业城一期住宅建筑立面设计
2008年，建筑面积18万m²
Architectural Facade Design of
Shum Yip's Community (Phase 1), Shunde
2008 Building area: 180000m²

中国饮食文化城建筑设计
2008年，用地面积220hm²，建筑面积38万m²
Architecture Design of
Chinese Gastrologic and Culture Town, Shenzhen
2008 Land area: 220ha
Building area: 380000m²

珠海五洲花城二期住宅建筑设计
2008年，用地面积16.5hm²，建筑面积60万m²
Architecture Design of
Five Continental Residence Garden, Zhuhai
2008 Land area: 16.5ha
Building area: 600000m²

深物业彩天怡色家园建筑设计
2008年，用地面积0.51hm²，建筑面积3.6万m²
Architecture Design of
'Colourful Sky' Residence, Shenzhen
2008 Land area: 0.51ha
Building area: 36000m²

成都麓山国际超高层住宅建筑设计
2008年，用地面积3.61hm²，建筑面积12万m²
Architecture Design of Super High-Rise Residential
Tower of Lushan International Community, Chengdu
2008 Land area: 3.61ha
Building area: 120000m²

成都天府华侨城二期住宅建筑设计
2008年，用地面积10hm²，建筑面积15万m²
Architecture Design of
Tianfu OCT Residence (Phase 2), Chengdu
2008 Land area: 10ha
Building area: 150000m²

中山市坦州镇居住小区住宅建筑设计
2008年，用地面积11.4hm²，建筑面积28万m²
Architecture Design of
Tanzhou Residential Community, Zhongshan
2008 Land area: 11.4ha
Building area: 280000m²

* 贵阳中天世纪新城三期四组团建筑设计
2007年，用地面积5.9hm²，建筑面积23.15万m²
Architecture Design of the 4# Block of
Zhongtian New Century City (Phase 3), Guiyang
2007 Land area: 5.9ha，
Building area: 231500m²

* 西安高科城市风景8#府邸建筑设计
2007年，用地面积2.6hm²，建筑面积9.10万m²
Architecture Design of
the 8# Mansion of Gaoke City View, Xi'an
2007 Land area: 2.6ha
Building area: 91000m²

深业·坪山居住小区规划与建筑方案设计
2007年，用地面积2.83hm²，建筑面积11.71万m²
Architecture Design of Pingshan
Residential Community
2007 Land area: 2.83ha，
Building area: 117100m²

珠海城市壹站居住区规划与建筑方案设计
2007年，用地面积2.85hm²，建筑面积8.31万m²
Architecture Design of The First City Station
Residential Community, Zhuhai
2007 Land area: 2.85ha
Building area: 83100m²

* 华润置地成都翡翠城四期住宅立面设计
2007年，用地面积7.56hm²，建筑面积41.56万m²
Architectural Facade Design of
CRL Jade-City Residence (Phase 4)
2007 Land area: 7.56ha，
Building area: 415600m²

佛山南海区狮山镇文化体育公园住宅设计
2007年，国际竞标，用地面积13.9hm²，建筑面积28万m²
Architecture Design of the Residence of Shishan
Culture& Sport Park, Nanhai district, Foshan
2007 International bidding Land area: 13.9ha
Building area: 280000m²

* 深圳市城建集团观澜居住区规划建筑设计
2007年，国际竞标中标方案，
用地面积16.99hm²，建筑面积34.6万m²
Architecture Design of Guanlan
Residential Community, Shenzhen
2007 Winning project of international bidding
Land area: 16.99ha Building area: 346000m²

* 鹏基惠州半山名苑居住建筑设计
2007年，用地面积2.6hm²，
建筑面积9.1万m²，竣工日期：2009年
Architecture Design of
Pengji Hillside Residential Community, Huizhou
2007 Land area: 2.60ha
Building area: 91000m² Constructed in 2009

江苏省姜堰市锦绣姜城规划与建筑方案设计
2007年，用地面积29.62hm²，
建筑面积62万m²
Architecture Design of Pengji
Splendid City, Jiangyan, Jiangsu Province
2007 Land area: 29.62ha
Building area: 620000m²

武汉万科高尔夫城市花园规划建筑设计
2007年，用地面积13.56hm²，
建筑面积27.9万m²
Architecture Design of
Vanke Golf City Garden Residence, Wuhan
2007 Land area: 13.56ha
Building area: 279000m²

深圳天鹅堡三期前期研究
2007年，用地面积11.17hm²，
建筑面积25.9万m²
Preliminary Studies for
The Swan Castle of OCT, Shenzhen
2007 Land area: 11.17ha
Building area: 259000m²

深圳中信惠州东江新城一期项目规划与建筑设计
2007年，用地面积25hm²，
建筑面积60万m²
Architecture Design of
CITIC Dongjiang New City (Phase 1), Huizhou
2007 Land area: 25ha
Building area: 600000m²

贵阳中天世纪新城三组团规划与建筑设计
2006年，用地面积17.54hm²，
建筑面积23.68万m²
Architecture Design of 3# Block of
Zhongtian New Century City, Guiyang
2006 Land area: 17.54ha
Building area: 236800m²

* 深圳蛇口君汇新天住宅小区规划与建筑设计
2006年，用地面积4.5hm²，
建筑面积11.65万m²
Architecture Design of
Junhuixintian Residential
Community, Shekou, Shenzhen
2006 Land area: 4.50ha
Building area: 116500m²

深圳市怡东花园规划建筑设计
2006年，用地面积6.1hm²，
建筑面积23.1万m²
Architecture Design of
Yidong Graden Residence, Shenzhen
2006 Land area: 6.10ha
Building area: 231000m²

* 合肥澜溪镇A、B区建筑及规划设计
2005年，用地面积7.23hm²，
建筑面积10.63万m²，
竣工日期：2007年，获奖项目
Planning Design of Section A&B of Nancy Town, Hefei
2005 Land area: 7.23ha Building area: 106300 m²
Constructed in 2007 Awarded project

深圳市宝安26区旧城改造项目住宅设计
2005年，国际竞标，用地面积23.13hm²，
建筑面积80.7万m²
Architecture Desig for Renovation of
26# Block in Bao'an, Shenzhen
2005 International bidding Land area: 23.13ha
Building area: 807000m²

招商华侨城尖岗山商业中心住宅设计
2005年，国际竞标，用地面积6.86hm²，
建筑面积6.53万m²
Architecture Design of
OCT Jiangganshan Shopping Center, Shen
2005 International bidding Land area: 6.
Building area: 65300m²

长沙天际岭项目住宅区设计
2005年，国际竞标，用地面积38.81hm²，
建筑面积39.77万m²
Architecture Design of
Tianjiling Residential Community, Changsha
2005 International bidding
Land area: 38.81ha Building area: 397700m²

苏州工业园区"左岸山庭"住宅区设计
2005年，国际竞标，用地面积19.54hm²，
建筑面积39.07万m²
Architecture Design of "Hill-Yard on Left-Bank"
Residence, Suzhou
2005 International bidding Land area: 19.54ha
Building area: 390700m²

成都阳明山庄城市别墅住宅区设计
2005年，用地面积33.34hm²，
建筑面积26.56万m²
Architecture Design of Yangming
Shanzhuang City Villa, Chengdu
2005 Land area: 33.34ha
Building area: 265600m²

* 成都天府长城2期住宅建筑设计
2005年，中标方案，用地面积6.22hm²，
建筑面积21万m²，竣工日期：2007年
Architecture Design of
Tianfu Great Wall Residence (Phase 2), Chen
2005 Winning project Land area: 6.22
Building area: 210000m² Constructed in 200

深圳市金光华"绿谷蓝溪"住宅小区设计
2005年，国际竞标，用地面积19.92hm²，
建筑面积35.84万m²
Architecture Design of Jinguanghua
"Lvgu Lanxi" Community, Shenzhen
2005 International bidding Land area: 19.92ha
Building area: 358400m²

北京密云休闲度假小区住宅设计
2005年，用地面积8hm²，
建筑面积14.28万m²
Architecture Design of
Beijing Miyun Vacation Residence
2005 Land area: 8ha
Building area: 142800m²

* 成都中海国际社区住宅设计
2004年，用地面积132.42hm²，
建筑面积396万m²
Design of Chengdu Zhonghai International
Community Residential
2004 Land area: 132.42ha
Building area: 3960000m²

* 华润置地成都翡翠城汇锦云天住宅建筑设计
2004年，国际竞标第二名，用地面积80hm²，
建筑面积86.65万m²，竣工日期：2007年
Architecture Design of
CRL Jade City Residence (Phase 2), Chen
2004, Second prize of international biddi
Land area: 80ha
Building area: 866500m² Constructed in 200

北京宣武区国信大吉片住宅设计
2004年，用地面积43.9hm²，
建筑面积165万m²
Planning and Design of
Xuanwu District Guoxin Dajipian Residence, Beijing
2004 Land area: 43.90ha
Building area: 1650000m²

湖南天健长沙芙蓉中路项目居住区设计
2004年，用地面积17.12hm²，
建筑面积75.12万m²
Architecture Design of
"Tianjian Changsha Furongzhong Road", Hunan
2004 Land area: 17.12ha,
Building area: 751200m²

福州市登云山庄居住区设计
2004年，用地面积267.4hm²，
建筑面积68.16万m²
Architecture Design of
Dengyun Mountain Village, Fuzhou
2004 Land area: 267.40ha
Building area: 681600m²

深圳市宝安城区26区商业公园居住区设计
2004年，规划面积26.7hm²，
建筑面积65万m²
Architecture Design of
Bao'an 26#Block Commercial Park, Shenzhen
2004 Planning area: 26.7ha
Building area: 650000m²

西安中海华庭居住小区设计
2004年，用地面积5.41hm²，
建筑面积17.3万m²
Architecture Design of
Zhonghai Luxuriant Garden Residence, Xi'an
2004 Land area: 5.41ha,
Building area: 173000m²

深圳市鸿荣源龙岗中心城居住区设计
2004年，国际竞标，用地面积40hm²，
建筑面积60万m²
Architecture Design of Hongrongyuan
Longgang Central City, Shenzhen
2004 International bidding Land area: 40ha,
Building area: 600000m²

深圳市西乡富通住宅区设计
2004年，总用地面积25hm²，
建筑面积50万m²
Architecture Design of Futong Residential
Community of Bao'an District, Shenzhen
2004 Land area: 25ha
Building area: 500000m²

上海金地普罗旺斯风情街住宅设计
2004年，用地面积2.3hm²，
建筑面积1.90万m²
Architecture Design of the Residence of
Jindi Provence Street, Shanghai
2004 Land area: 2.30ha
Building area: 19000m²

* 深圳市恒立海岸花园居住区设计
2004年，用地面积2hm²，
建筑面积6万m²，竣工日期：2005年
Architecture Design of
Hengli Seashore Garden Residence, Shenzhen
2004 Land area: 2ha
Building area: 60000m² Constructed in 2005

深圳天琴湾项目三期别墅区单体设计
2004年，用地面积0.5hm²，
建筑面积1720m²
Architecture Design of
Lyra Bay Villa (Phase 3), Shenzhen
2004 Land area: 0.5ha,
Building area: 1720m²

深圳市沙河世纪山谷住宅小区设计
2003年，用地面积18hm²，
建筑面积67万m²
Architecture Design of
Shahe Century Valley Residential Community, Shenzhen
2003 Land area: 18ha
Building area: 670000m²

苏州新加坡工业园区东湖大郡二期住宅小区设计
2003年，用地面积10hm²，
建筑面积18万m²
Architecture Design of East Lake Residence
(Phase 2) of Singapore Industry Park, Suzhou
2003 Land area: 10ha
Building area: 180000m²

东莞世纪新城首期住宅小区设计
2003年，用地面积13hm²，
建筑面积18万m²
Architecture Design of
Center New City (Phase 1), Dongguan
2003 Land area: 13ha
Building area: 180000m²

* 苏州工业园区东城郡住宅建筑设计
2003年，用地面积6.5hm²，
建筑面积14.70万m²，竣工日期：2005年
Architecture Design of Residential Community of
East County of Suzhou Industry Park
2003 Land area: 6.50ha
Building area: 147000m² Constructed in 2005

上海西郊古北国际别墅设计
2003年，建筑面积602.32m²
Architecture Design of
Gubei International Villa, Shanghai
2003 Building area: 602.32m²

* 成都新城市广场公寓设计
2002年，建筑面积17万m²，
竣工日期：2005年
Architecture Design of
New City Plaza Apartment, Chengdu
2002 Building area: 170000m²,
Constructed in 2005

* 杭州盛德嘉苑三期住宅小区设计
2002年，建筑面积6.1万m²，
竣工日期：2005年
Architecture Design of
Shengdejiayuan (Phase 3), Hangzhou
2002 Building area: 61000m²,
Constructed in 2005

* 深圳半岛城邦1期住宅建筑设计
2002年，用地面积5hm²，建筑面积18万m²，
竣工日期：2007年 获奖项目
Architecture Design of
The Peninsula Residence (Phase 1), Shenzhen
2002 Land area: 5ha Building area: 180000m²
Constructed in 2007 Award project

上海嘉定高尔夫居住区设计
2002年，用地面积667hm²
Architecture Design of
Jiading Golf Residential Community, Shanghai
2002 Land area: 667ha

南京将军山别墅及多层住宅小区设计
2002年，用地面积16hm²，
建筑面积23.50万m²
Architecture Design of
Mt.General Villa and Multi-
floored Residence, Nanjing
2002 Land area: 16ha Building area: 235000m²

* 福州融侨"江南水都"首期住宅建筑设计
2002年，中标方案，用地面积12.34hm²，
建筑面积18.4万m²，竣工日期：2006年
Architecture Design of Rongqiao
"Jiang nan Water Town" Residence (Phase 1), Fuzhou
2002 Winning project Land area: 12.34ha
Building area: 184000m² Constructed in 2006

深圳国际网球中心俱乐部小区建筑设计
2001年，用地面积9.8hm²，住宅建筑面积14万m²，
国际网球中心建筑面积3万m²
Architecture Design of International
Tennis Center Club Residence, Shenzhen
2001 Land area: 9.8ha
Building area of house: 140000m², Building
area of international tennis center: 30000m²

* 贵阳中天世纪新城一期联排住宅小区建筑设计
2001年，用地面积8.4hm²，
建筑面积5.2万m²，竣工日期：2003年
Architecture Design of the Townhouse of
Zhongtian Century New City (Phase 1), Guiyang
2001 Land area: 8.40ha
Building area: 52000m² Constructed in 2003

成都长城地产五洲花园住宅区设计
2001年，用地面积63hm²，
建筑面积80万m²
Architecture Design of Changcheng Five
Continents Residence Garden, Chengdu
2001 Land area: 63ha
Building area: 800000m²

深圳万科下沙滨海住宅设计
2001年，建筑面积24万m²
Architecture Design of
Vanke Xisha Seashore Residence, Shenzhen
2001 Building area: 240000m²

深圳星河地产"叠翠九重天"住宅设计
2001年，用地面积59.71hm²，
建筑面积40万m²
Architecture Design of
Xinghe Real Estate "Rich Green Paradise"
Residential Community, Shenzhen
2001 Land area: 59.71ha
Building area: 400000m²

北京万科青青家园住宅小区设计
2000年，用地面积23hm²，
建筑面积28万m²
Architecture Design of
Vanke Qingqing House, Beijing
2000 Land area: 23ha
Building area: 280000m²

深圳华侨城中央花园住宅设计
2000年，建筑面积2.8万m²
Architecture Design of OCT Central Garden
Residence, Shenzhen
2000 Building area: 28000m²

杭州市吴山商城住宅设计
1999年，国际竞标，用地面积6hm²，
建筑面积9万m²
Architecture Design of the Residence of
Wushan Business City
1999 International bidding
Land area: 6ha Building area: 90000m²

山东省青岛世纪广场住宅设计
1999年，用地面积3hm²，
建筑面积6万m²
Architecture Design of
Century Plaza Residence, Qingdao
1999 Land area: 3ha
Building area: 60000m²

* 深圳盐田碧海名峰（现名天琴海）别墅区住宅设计
1998年，用地面积9hm²，
建筑面积3.8万m²
Architecture Design of Bihaimingfeng
(Lyra Bay) Villa, Yiantian District, Shenzhen
1998 Land area: 9ha
Building area: 38000m²

* 深圳世界花园第八期—海华居住宅设计
1998年，建筑面积15万m²，
竣工日期：2001年
Architecture Design of World Garden (Phase 8)
—Haihua Residence, Shenzhen
1998 Building area: 150000m²
Constructed in 2001

深圳市世界广场住宅设计
1998年，国际竞标中标方案，
建筑面积10.5万m²
Architecture Design of World Plaza Residence,
Shenzhen
1998 Winning project of international bidding
Building area: 105000m²

深圳观澜高尔夫球会会员住宅设计
1998年，建筑面积7.7万m²
Architecture Design of Guanlan Golf Club
VIP's Residence, Shenzhen
1998 Building area: 77000m²

深圳雅庭苑住宅小区建筑设计
1998年，国际竞标中标方案，
建筑面积10.4万m²
Architecture Design of Yating Residence, Shenzhen
1998 Winning project of international bidding
Building area: 104000m²

江苏省无锡法国达能矿泉水厂和生活基地设计
1997年，建筑面积3万m²
Architecture Design of French Daneng Mineral
Water Factory and Living Area, Wuxi, Jiangsu
1997 Building area: 30000m²

广西壮族自治区来宾法国电力公司厂区住宅设计
1997年，建筑面积3万m²
Architecture Design of Residential Area of Laibin
French Power Company Workshop, Guangxi
1997 Building area: 30000m²

江苏东海市圣戈班玻璃制厂和生活基地法国专家村设计
1997年，用地面积18hm²，建筑面积2.5万m²
Architecture Design of St. Gorban Glass
Manufacturer and Living Area French Expert
Village, Donghai, Jiangsu
1997 Land area: 18ha
Building area: 25000m²

北京国门广场住宅区设计
1995年，建筑面积28万m²
Architecture Design of Beijing National Gate Plaza
Residential Community
1995 Building area: 280000m²

上海市浦东六里现代化生活居住园区设计
1995年，建筑面积30万m²
Architecture Design of Pudong Six Li
Modern Life Residence Community, Shanghai
1995 Building area: 300000m²

景观
Landscape

项目总数: 117
已竣工或建造中项目: 52

* 已竣工或建造中项目
* Projects constructed or under construction

* 贵阳云岩区十里花川公共空间及南明河两侧景观概念设计
2010年，委托设计，用地面积 96.7 hm²，
景观面积63.4万m²
Landscape Design of 10 Miles Flower Valley of Yu'an-Anjing District, Yunyan, Guiyang
2010 Mandated project Land area: 96.7 ha
Landscape area: 634 000 m²

* 昆明滇越铁路主题公园景观方案设计
2010年，委托设计，用地面积 52 hm²，
景观面积40万m²
Landscape Design of French New Station Park of Dian Lake of Kunming
2010 Mandated project Land area: 100ha
Landscape area: 400,000m²

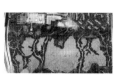

* 贵阳国际会议展览中心西侧绿地公园景观设计
2010年，委托设计，用地面积 24 hm²，
景观面积24万m²
Landscape Design of Green Land Park on the west of Guiyang International Conference & Exhibition Center
2010 Mandated project Land area: 24 ha

* 深圳康佳研发大厦景观设计
2010年，委托设计，用地面积 0.96 hm²，
景观面积1.12万m²
Landscape Design of Konka R&D Tower of Shenzhen
2010 Mandated project Land area: 0.96 ha
Landscape area: 11 200 m²

* 深圳长富金茂大厦景观设计
2010年，委托设计，用地面积 1.88 hm²，
景观面积1.37万m²
Landscape Design of Finance Tower of Changfu of Shenzhen
2010 Mandated project Land area: 1.88 ha
Landscape area: 13700 m²

* 深圳深业新岸线入口及商业街景观改造设计
2010年，委托设计，用地面积 1.6 hm²，
景观面积1.58万m²
Landscape Design of the Entry and Retail Street of Shum-yip New Shoreline Community of Shenzhen
2010 Mandated project Land area: 1.6 ha
Building area: 15 800m²

* 贵阳中天世纪新城6、7组团景观设计
2010年，委托设计，用地面积 10.4 hm²，
景观面积4.88万m²
Landscape Design of Blocks 6&7# of New Century City of Guiyang
2010 Mandated project Land area: 10.4 ha
Landscape area: 48 800 m²

* 贵阳中天世纪新城4组团幼儿园景观设计
2010年，委托设计，用地面积0.44 hm²，
景观面积0.44万m²
Landscape Design for the Kindergarden of
Block 4 of New Century City of Guiyang
2010 Mandated project Land area: 0.44 ha
Landscape area: 4400 m²

* 福州市闽江北岸中央商务中心城市广场景观设计
2010年，竞标方案，景观面积5.6万m²
Landscape Design of City Plaza of CBD
Northern Bank of Minjiang River, Fuzhou
2010 Bidding project
Landscape area: 56 000 m²

* 贵阳国际会议展览中心景观设计
2009年，委托设计，用地面积51.8hm²，景观面积31.81hm²
Landscape Design of Guiyang International
Conference & Exposition Center
2009 Mandated project Land area: 51.8ha
landscape area 318100m²

深圳南头古城——2街1园景观设计
2009年，竞标方案，用地面积10.8hm²
Landscape Design of Shenzhen Nantou Ancient
City - 2 streets & 1 park
2009 Bidding project Land area: 10.8ha

泉州洛江现代世界农业旅游观光生态小镇
景观概念深化设计
2009年，委托设计，用地面积510.8hm²
Conceptional Design of Ecological Town for
Luojiang Modern Agricultural Tourism, Quanzhou
2009 Mandated project Land area: 510.8ha

* 深圳长富金茂大厦景观设计
2009年，委托设计，用地面积1.88hm²
Landscape Design of Changfu Jinmao Tower, Shenzhen
2009 Mandated project Land area: 1.88ha

* 中山金源花园景观设计
2009年，中标方案，用地面积21.22hm²
Landscape Design of Jin Yuan Garden
Residence in Zhongshan
2009 Winning project Land area: 21.22ha

* 深圳城建集团碧中园二期景观设计
2009年，中标方案，用地面积2.5hm²，
景观面积2.2万m²
Landscape Design of SZEG Bi Zhong Yuan
Residence (phase II) in Shenzhen
2009 Winning project Land area: 2.5ha
landscape area 22000m²

* 深圳城建集团观澜项目景观设计
2009年，中标方案，用地面积20hm²，
景观面积13万m²
Architectural Design of SZEG Guanlan Residential
Community in Shenzhen
2009 Winning project Land area: 20ha
landsacpe area:130000m²

* 深圳半岛城邦三期景观设计
2009年，委托设计，用地面积5.68hm²
Landscape Design of Shenzhen Peninsula (phase III)
2009 Mandated project Land area: 5.68ha

* 昆明滇池法国新站主题公园景观设计
2009年，竞标方案，用地面积100hm²，
景观公园面积40万m²
Landscape Design of French New Station of Dian
Lake of Kunming
2009 Bidding project Land area: 100 ha
landscape area: 400000m²

深圳市宝安中心区银晖路商业街及
核心商业区步行商业街景观设计
2008年，中标方案，用地面积4.95hm²
Landscape Design of Yinhui Shopping Street and
Pedestrian Shopping Street of
Core Shopping Area in
Bao'an Central District, Shenzhen
2008 Winning project Land area: 4.95ha

深圳人才园景观设计
2008年，用地面积3.61hm²
Landscape Design of
Human Resoucres Park, Shenzhen
2008 Land area: 3.61ha

* 深圳北站交通枢纽工程景观设计
2008年，中标方案，
用地面积10hm²，
Landscape Design of Transportation Hub of
North Railway Station, Shenzhen
2008 Winning project
Land area: 10ha

珠海歌剧院景观设计
2008年，用地面积42hm²
Landscape Design of Zhuhai Opera House
2008 Land area: 42ha

成都怡湖玫瑰湾景观设计
2008年，用地面积12.57hm²
Landscape Design of Yihu Rose Bay Residential
Community, Chengdu
2008 Land area: 12.57ha

深圳市光明新区公明文化艺术和体育中心景观设计
2008年，用地面积9.9hm²
Landscape Design of Gongming Culture, Art &
Sports Center of New Guangming Town, Shenzhen
2008 Land area: 9.9ha

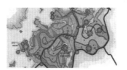

泉州洛江农业旅游生态观光小镇景观概念设计
2008年，用地面积265hm²
Conceptual Landscape Design of Luojiang Ecological &
Agricultural Tourist Town, Quanzhou
2008 Land area: 265ha

* 贵阳中天世纪新城中心商业组团景观设计
2008年，用地面积8.6hm²
Landscape Design of Central Commercial
Block of Zhongtian Century New Town, Guiyang
2008 Land area: 8.6ha

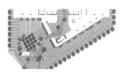

* 深圳康佳集团研发大厦景观设计
2008年，中标方案，
用地面积0.96hm²
Landscape Design of
Konka R&D Building, Shenzhen
2008 Winning project
Land area: 0.96ha

晋江新中心区市民广场景观概念设计
2008年，用地面积11hm²
Conceptual Landscape Design of
Citizen Plaza of New City Center of Jinjiang
2008 Land area: 11ha

* 成都中航城市广场景观设计
2008年，用地面积1.98hm²
Landscape Desgin of
CATIC City Plaza, Chengdu
2008 Land area: 1.98ha

成都中航国际广场景观设计
2008年，用地面积1.5hm²
Landscape Design of
CATIC International Plaza, Chengdu
2008 Land area: 1.5ha

中国饮食文化城景观设计
2008年，用地面积220hm²
Landscape Desgin of
Chinese Gastrologic and Culture Town, Shenz
2008 Land area: 220ha

深圳市南油购物公园景观设计
2008年，国际竞标第一名，
用地面积13hm²
Landscape Design of
Nanyou Shopping Park, Shenzhen
2008 First prize of International bidding
Land area: 13ha

珠海五洲花城二期概念性景观设计
2008年，用地面积16.5hm²
Landscape Design of
Five Continental Residence Garden, Zhuhai
2008 Land area: 16.5ha

* 深圳宝安区石岩大树林体育公园景观设计
2008年，中标方案
用地面积5.1hm²
Landscape Design of
Shiyan Sporting Park of Bao'an District, Shenzhen
2008 Winning project
Land area: 5.1ha

深圳宝安区石岩水源保护公园景观设计
2008年，中标方案，
用地面积26.7hm²
Landscape Design of
Shiyan Water Resource Protection Park of B
District, Shenzhen
2008 Winning project
Land area: 26.7ha

深圳南澳月亮湾海岸带景观改造概念规划设计
2008年，用地面积14.2hm²
Conceptual Landscape Plannning of
Moon Bay Coast of Nan'ao, Shenzhen
2008 Land area: 14.2ha

深圳市光明新区中央公园景观概念规划设计
2008年，用地面积237hm²
Conceptual Landscape Planning of Central Park
of Guangming New Town, Shenzhen
2008 Land area: 237ha

* 长沙楚家湖景观设计
2007年，用地面积62.5万m²，
景观面积：46.88万m²
Landscape Design of Chujia Lake Residential
Community, Changsha
2007 Land area: 62.5ha
Landscape area: 468800m²

* 花样年(成都)蒲江大溪谷一期景观设计
2007年，用地面积17hm²，
景观面积12.05万m²，竣工日期：2009年
Landscape Design of Pujiang Grand Valle
Fantasia Group (Phase 1), Chengdu
2007 Land area: 17ha
Landscape area: 120500m² Constructed in 2

* 深圳三湘国际花园景观设计
2007年，用地面积9.27hm²，
景观面积7万m²，竣工日期：2010年
Landscape Design of
Sanxiang International Garden Residence, Shenzhen
2007 Land area: 9.27ha
Landscape area: 70000m² Constructed in 2010

* 深圳龙岗城龙花园四期景观设计
2007年，用地面积3.5hm²，
景观面积3万m²
Landscape Design of Chenglong
Garden (Phase 4), Longgang, Shenzhen
2007 Land area: 3.5ha
Landscape area: 30000m²

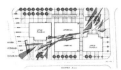
* 深圳龙光世纪大厦景观概念设计
2007年，用地面积1.6hm²，
景观面积1.6万m²
Landscape Design of
Longguang Century Tower, Shenzhen
2007 Land area: 1.60ha
Landscape area: 16000m²

* 万科宁波金色水岸居住区景观设计
2007年，用地面积13hm²，
景观面积6万m²
Landscape Design of Vanke Golden Waterfron
Residential Community, Ningbo
2007 Land area: 13ha
Landscape area: 60000m²

* 深圳金光华春华四季园二期景观设计
2007年，用地面积27hm²，
景观面积8万m²
Landscape Design of Jinguanghua"Chunhua
FourSeasons Garden" (Phase 2), Shenzhen
2007 Land area: 27ha
Landscape area: 80000m²

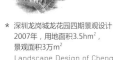
深圳市罗湖区贝丽中学(水贝珠宝学校)景观设计
2007年，国际竞标，
用地面积1.95hm²
Landscape Design of
Beili Technical School, Shenzhen
2007 International bidding
Land area: 1.95ha

深圳蛇口君汇新天住宅小区景观设计
2007年，用地面积4.5hm²，
竣工日期：2009年
Landscape Design of Junhuixintian Residential
Community, Sekou, Shenzhen
2007 Land area: 4.5tha
Constructed in 2009

* 深圳半岛城邦二期销售中心景观设计
2007年，用地面积2800m²，
景观面积2800m²
Landscape Design of Sales Center of
The Peninsula (Phase 2), Shenzhen
2007 Land area: 2800m²
Landscape area: 2800m²

佛山南海区狮山镇文化体育公园景观设计
2007年，国际竞标，
用地面积13.9hm²
Landscape Design of Shishan Culture &
Sport Park, Nanhai district, Foshan
2007 International bidding
Land area: 13.9ha

珠海前山新冲路城市壹站景观设计
2007年，国际竞标，
用地面积2.85hm²
Landscape Design of the First City Station, Zhuhai
2007 International bidding
Land area: 2.85ha

深圳市龙岗区"深业·坪山"居住区景观设计
2007年，国际竞标，
用地面积2.83hm²
Landscape Design of "Shum Yip Pingshan"
Residential Community, Longang district, Shenzhen
2007 International bidding
Land area: 2.83ha

F1摩托艇深圳赛区景观设计
2007年，国际竞标
用地面积70hm²，景观面积70万m²
Landscape Design of Shenzhen Station of
F1 Powerboat World Championship
2007 International bidding
Land area: 70ha Landscape area: 700000m²

武汉万科高尔夫城市花园景观设计
2007年，用地面积12.56hm²，
景观面积34.3万m²
Landscape Design of
Vanlke Golf City Garden Residence, Wuhan
2007 Land area: 12.56ha
Landscape area: 343000m²

★ 鹏基惠州半山名苑景观设计
2007年，用地面积49.71hm²，
景观面积10.78万m²，竣工日期：2009年
Landscape Design of
Pengji Hillside Residential Community, Huizhou
2007 Land area: 49.71ha
Landscape area: 107800m² Constructed in 2009

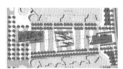
★ 深圳南山海德二道海印长城段景观设计
2007年，用地面积1hm²
竣工日期：2008年
Landscape Design of Haide'er Road
(Haiyin Changcheng Part), Shenzhen
2007 Land area: 1ha
Constructed in 2008

★ 深圳蓝郡广场景观设计
2007年，用地面积18hm²
竣工日期：2008年
Landscape Design of Lanjun Plaza, Shenzhen
2007 Land area: 18ha
Constructed in 2008

★ 深圳沙河世纪假日广场景观设计
2007年，用地面积1.63hm²，
景观面积1.3万m²，竣工日期：2009年
Landscape Design of
Shahe Century & Holiday Plaza, Shenzhen
2007 Land area: 1.63ha
Landscape area: 13000m² Constructed in 2009

★ 西安高科城市风景8#府邸景观设计
2007年，用地面积12.56hm²，
景观面积34.3万m²
Landscape Design of
the 8# Mansion of Gaoke City View, Xi'an
2007 Land area: 12.56ha
Landscape area: 343000m²

深圳华侨城"欢乐海岸"总体规划
2006年，用地面积56.46hm²
General Planning of Shenzhen OCT Happy Coast
2006 Land area: 56.46ha

成都市青白江区湿地公园方案设计
2006年，用地面积55hm²
Landscape Design of
Wetland Park of QingBaiJiang District, ChengDu
2006 Land area: 55ha

中央电视台媒体公园景观设计
2006年，国际竞标，
用地面积2.56hm²
Landscape Design of CCTV Media Park
2006 International bidding
Land area: 2.56ha

苏州太湖国家旅游度假区入口节点景观设计
2006年，国际竞标中标方案，
用地面积101400hm²
Landscape Design of Entrance Area of
Taihu Lake National Tourist Zone, Suzhou
2006 Winning project of international bidding
Land area: 101400ha

★ 成都天府长城2期景观设计
2005年，用地面积5.14hm²，
竣工日期：2007年
Landscape Design of
Tianfu Great Wall Residence (Phase2), Chengdu
2005 Land area: 5.14ha
Constructed in 2007

深圳市蛇口微波山顶景观设计
2005年，用地面积1100m²
Landscape Design of
the Top of Weibo Hill, Shekou, Shenzhen
2005 Land area: 1100m²

★ 半岛城邦一期居住区及滨海带景观设计
2005年，用地面积8.36hm²
竣工日期：2007年 获奖项目
Landscape Design of The Peninsula (Phase 1)
and the Seashore Landscape Belt, Shenzhen
2005 Land area: 8.36ha
Constructed in 2007 Awarded project

★ 深圳市人民南路环境景观改造方案设计
2004年，用地面积12.44hm²，
竣工日期：2008年 获奖项目
Landscape Design of
Renminnan Boulvard, Shenzhen
2004 Land area: 12.44ha
Constructed in 2008 Awarded project

深圳市沙河世纪山谷住宅小区景观设计
2004年，用地面积18hm²
Landscape Design of
Shahe Century Valley Residence, Shenzhen
2004 Land area: 18ha

深圳市鸿荣源龙岗中心城景观设计
2004年，国际竞标，
用地面积40hm²
Landscape Design of
Hong Rongyuan Longgang Central City, Shenzhen
2004 International bidding
Land area: 40ha

* 成都华润置地.翡翠城二期及府河公园景观设计
2004年，国际竞标第二名，
用地面积12.70hm², 景观面积2.2万m²
Landscape Design of CRL Jade City (Phase 2) and Fuhe Park, Chengdu
2004 Second prize of international bidding
Land area: 12.7ha Landscape area: 22000m²

* 深圳金光华春华四季园景观设计
2004年，用地面积27hm²,
景观面积8万m² 竣工日期: 2007年
Landscape Design of Jinguanghua "Chunhua FourSeasons Garden", Shenzhen
2004 Land area: 27ha
Landscape area: 80000m² Constructed in 2007

西安中海华庭居住小区景观设计
2004年，用地面积5.41hm²
Landscape Design of
Zhonghai Huating Residential Community, Xi'an
2004 Land area: 5.41ha

* 深圳市南山区商业文化中心区环境设计
2004年，国际竞标中标方案，
用地面积30hm² 竣工日期: 2009年
Landscape Design of Commercial & Cultural Center of Nanshan, Shenzhen
2004 Winning project of international bidding
Land area: 30ha Constructed in 2009

福州市登云山庄景观设计
2004年，用地面积267.4hm²
Landscape Design of
Dengyun Mountain Village, Fuzhou
2004 Land area: 267.4ha

湖南天健长沙芙蓉中路项目居住区景观设计
2004年，用地面积17.12hm²
Landscape Design of
"Tianjian Changsha Furongzhong Road", Hunan
2004 Land area: 17.12ha

深圳盐田中心区城市和中轴线景观设计
2003年，城市规划设计面积18hm²,
环境景观设计面积6.5hm²
Landscape Design of the Center and Central Axis of Yantian District, Shenzhen
2003 Urban planning area: 18ha
Landscape design area: 6.5ha

* 深圳大剧院环境景观改造设计
2003年，景观面积3hm²,
竣工日期: 2006年
Landscape Design of Shenzhen Grand Theatre
2003 Landscape area: 3ha
Constructed in 2006

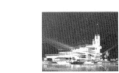

深圳蛇口东填海区大型住宅区景观设计
2003年，用地面积325hm²
Landscape Design of the Large Scale Residential Community of East Reclamation Area of Shekou, Shenzhen
2003 Land area: 325ha

海南省三亚阳光海岸景观设计
2003年，国际竞标，
设计面积146.67hm²
Landscape Design of Sanya Sunny-Coast, Hainan
2003 International bidding
Design area: 146.67ha

深圳湾滨海带景观设计
2003年，国际竞标第二名，
设计范围长15公里
Landscape Design of Sea Belt of Shenzhen Bay
2003 Second prize of international bidding
Landscape design scope: 15km

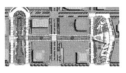

珠海情侣路滨海带景观设计
2003年，国际竞标，海岸线6.5公里
规划环境景观面积150hm²
Landscape Design of lovers road of Zhuhai
2003 International bidding Coastline: 6.5km
Planning & landscape design area: 150ha

深圳宝安国际机场入口景观设计
2003年，国际竞标，
用地面积30hm²
Landscape Design of Bao'an International Airport Entrances, Shenzhen
2003 International bidding
Land area: 30ha

* 重庆融侨半岛一期2#地块B3区景观设计
2003年，景观设计面积1.75hm²,
竣工日期: 2004年
Landscape Design of B3 Plot of Block 2 of Rongqiao Peninsula (Phase 1), Chongqing
2003 Landscape area: 1.75ha
Constructed in 2004

苏州新加坡工业园区东湖大郡二期住宅小区景观设计
2003年，用地面积10hm²
Landscape Design of East Lake Residence (Phase 2) of Singapore Industry Park, Suzhou
2003 Land area: 10ha

深圳市中心区22, 23-1街坊街道设施和公园景观设计
2003年，用地面积17hm²
Landscape Design of Streetscape and Park Block 22,23-1 of Central Area, Shenzhen
2003 Land area: 17ha

* 广州玖玖文化家园景观设计
2003年，用地面积4.37hm²,
景观面积3.5万m², 竣工日期: 2004年
Landscape Design of
Jiujiu Culture Home, Guangzhou
2003 Land area: 4.37ha
Landscape area: 35000m² Constructed in 2004

深圳市华侨城玫瑰广场景观设计
2003年，用地面积1.6hm²
Landscape Design of
OCT Rose Plaza, Shenzhen
2003 Land area: 1.6ha

中央音乐学院珠海分校景观设计
2003年，国际竞标，用地面积43hm²
Landscape Design of
National Music University-Zhuhai Branch
2003 International bidding Land area: 43ha

苏州工业园区东城郡景观设计
2003年，用地面积10hm²,
竣工日期: 2006年
Landscape Design of East County Residential Community of Suzhou Industry Park
2003 Land area: 10ha
Constructed in 2006

重庆融侨半岛一期2#地块B1，B2区景观设计
2002年，用地面积11.7hm²，
竣工日期：2004年
Landscape Design of B1& B2 Plots of Block 2 of
Rongqiao Peninsula (Phase 1), Chongqing
2002 Land area: 11.7ha
Constructed in 2004

＊ 深圳市龙岗下沙金沙湾休闲区景观设计
2002年，用地面积17hm²，
竣工日期：2003年
Landscape Design of
Golden-Beach Leisure Area of Xiasha, Shenzhen
2002 Land area: 17ha
Constructed in 2003

上海嘉定高尔夫社区景观设计
2002年，用地面积667hm²
Landscape Design of
Jiading Golf Community, Shanghai
2002 Land area: 667ha

＊ 深圳中海阳光棕榈园二期景观设计
2002年，景观面积3.3hm²，
竣工日期：2003年
Landscape Design of
Zhonghai Sunny Palm Garden (Phase 2), Shenzhen
2002 Landscape area: 3.30ha
Constructed in 2003

＊ 广州瑞丰中大花园环境景观设计
2002年，用地面积33.69hm²，
景观面积1.1万平方米，竣工日期：2003年
Landscape Design of
Ruifeng Zhongda Garden, Guangzhou
2002 Land area: 33.69ha
Landscape area: 11000 m² Constructed in 2003

＊ 深圳市华侨城学校景观设计
2002年，设计面积2hm²，
竣工日期：2003年
Landscape Design of OCT School, Shenzhen
2002 Design area: 2ha
Constructed in 2003

桂林市临桂县九里香堤别墅环境景观设计
2002年，景观面积12hm²
Landscape Design of Jiuli
Frangarant Mound Villa of Lingui, Guilin
2002 Landscape area: 12ha

深圳松泉山庄三期环境景观设计
2002年，用地面积1.92hm²，
环境景观用地面积1.6万m²
Landscape Design of Pine Tree and
Spring Mountain Village (Phase 3), Shenzhen
2002 Land area: 1.92ha
Landscape area: 16000m²

＊ 福州融侨锦江A～二期环境景观设计
2002年，景观面积1hm²，
竣工日期：2004年
Landscape Design of
Rongqiao Jinjiang A - Phase 2, Fuzhou
2002 Landscape area: 1ha
Constructed in 2004

济南市南部新城区城市中心景观设计
2001年，规划面积396hm²
Landscape Design of
the Center of New South District, Jinan
2001 Planning area: 396ha

杭州市钱塘江两岸新市中心区景观设计
2001年，国际竞标，
用地面积300hm²
Landscape Design of
New City Centers along Qiantang River, Hangzhou
2001 International bidding
Land area: 300ha

深圳星河地产"叠翠九重天"住宅景观设计
2001年，用地面积59.71hm²
Landscape Design of
Xinghe Real Estate "Rich Green Paradise"
Residential Community, Shenzhen
2001 Land area: 59.71ha

成都长城地产五洲花园住宅区景观设计
2001年，国际竞标，
用地面积63hm²
Landscape Design of Changcheng Five
Continents Residence Garden, Chengdu
2001 International bidding
Land area: 63ha

吉林市松花江沿岸总体规划构思及重点地区城市景观设计
2001年，国际竞标，规划陆地面积2525hm²，
城市设计范围971hm²
General Planining of Songhua River shores
and Landscape Design for the Key Areas, Jilin
2001 International bidding Land area: 2525ha
Urban Design area: 971ha

深圳市智泉苑景观设计
2001年，设计面积0.7hm²
Landscape Design of
Zhiquanyuan Residence, Shenzhen
2001 Design area: 0.7ha

南海市海八路以北景观设计
2001年，用地面积174hm²
Landscape Design for the North of
Haiba Road, Nanhai
2001 Land area: 174ha

＊ 重庆华立天地豪园景观设计
2001年，环境景观用地面积2.5hm²，
竣工日期：2003年
Landscape Design of
Huali World Luxuriant Garden, Chongqing
2001 Landscape area: 2.5ha
Constructed in 2003

＊ 广州新世界地产凯旋新世界景观设计
2001年，景观面积10hm²，
竣工日期：2004年
Landscape Design of
Triumph New-World Garden, Guangzhou
2001 Landscape area: 10ha
Constructed in 2004

深圳龙岗区体育文化广场景观设计
2001年，设计面积2.5hm²
Landscape Design of
Longgang Gym and Culture Square, Shenzhen
2001 Design area: 2.5ha

深圳高新技术开发区中心区景观设计
2001年，用地面积25.6hm²
Landscape Design for the Central Part of
Shenzhen Hi-tech Development Zone
2001 Land area: 25.6ha

深圳国际网球中心俱乐部小区景观设计
2001年，用地面积9.8hm²
Landscape Design of International
Tennis Center Club Residence, Shenzhen
2001 Land area: 9.8ha

* 深圳中海深圳湾畔花园景观设计
2001年，用地面积2.5hm²，
竣工日期：2002年
Landscape Design of
Zhonghai Seaview Garden, Shenzhen
2001 Land area: 2.5ha
Constructed in 2002

湖州市仁皇山新区城市景观设计
2001年，国际竞标，
用地面积400hm²
Landscape Design of
Renhuangshan New District, Huzhou
2001 International bidding
Land area: 400ha

* 福州融侨水乡温泉别墅景观设计，
2001年，环境景观用地面积11m²，
竣工日期：2003年
Landscape Design of
Hot Spring Waterfront Villas of Rongqiao, Fuz
2001 Landscape area: 11ha
Constructed in 2003

温州苍南新市中心区景观设计
2001年，用地面积454m²
Landscape Design of
New City Center of Cangnan, Wenzhou
2001 Land area: 454ha

* 深圳深南中路环境及灯光设计
2000年，5公里长，用地面积90hm²，
竣工日期：2000年
Landscape and Lighting Design of
Shennanzhong Road, Shenzhen
2000 Length:5km Land area: 90ha
Constructed in 2000

深圳市中心公园红荔路以北景观设计
2000年，设计面积36hm²
Landscape Design of Shenzhen Central Park
2000, Design area: 36ha

* 安徽合肥国际会展中心景观设计
2000年，国际竞标中标方案，
用地面积8.30hm²，竣工日期：2002年
Landscape Design of Hefei International
Conference & Exhibition Center, Anhui
2000 Winning project of international bid
Land area: 8.30ha Constructed in 2002

深圳宝安中心城环境设计
2000年，国际竞标，
用地面积100hm²，绿地面积30万m²
Landscpel Design of
Bao'an Central City, Shenzhen
2000 International bidding
Land area: 100ha Green area: 300000m²

* 深圳市宝安区湖景居环境设计
2000年，用地面积0.7hm²，
竣工日期：2000年
Landscpe Design of Lake View Residence of
Bao'an District, Shenzhen
2000 Land area: 0.7ha Constructed in 2000

深圳市城市入口公园景观设计
2000年，用地面积3.5hm²
Landscape Design of
City Entrance Park, Shenzhen
2000 Land area: 3.5ha

* 深圳市盐田区沙头角综合体育中心景观设计
2000年，用地面积0.46hm²，
竣工日期：2004年
Landscape Design of
Yantian Shatoujiao Sports Center, Shenzhe
2000 Land area: 0.46ha Constructed in 2

深圳莲花山广场环境景观概念设计
2000年，设计面积1.5m²
Conceptual Landscape Design of
Lianhuashan Square, Shenzhen
2000 Design area: 1.5ha

山东省青岛世纪广场景观设计
1999年，用地面积3hm²
Landscape Design of
Shandong Qingdao Century Square
1999 Land area: 3ha

* 深圳银湖度假区棕榈泉环境景观设计
1999年，用地面积1.50hm²，
竣工日期：2000年
Landscape Design of Palm Spa Area of
Silver Lake Holiday Zone, Shenzhen
1999 Land area: 1.50ha Constructed in 2000

* 深圳规划国地局盐田分局办公楼景观设计
1999年，国际竞标中标方案，
用地面积，0.6hm²，竣工日期：2001年
Landscape Design of Yantian Office Building
Shenzhen Urban Planning & Land Bureau
1999 Winning project of international bidd
Land area: 0.6ha Constructed in 2001

* 深圳时钟广场景观设计
1999年，市政府委托，
样板工程，竣工日期：2002年
Landscape Design of Clock-plaza, Shenzhen
1999, Entrusted by Shenzhen Government,
Model project Constructed in 2002

广西桂林市水系景观设计
1999年，国际竞标，
核心范围300hm²，影响范围600hm²
Landscape Design of Guilin Watersystem, Guangxi
1999 International bidding
Core area 300ha Influence area 600h

* 深圳华侨城OCT生态广场景观设计
1998年，用地面积5hm²，竣工日期：1999年
Landscape Design of
OCT Ecology Plaza, Shenzhen
1998 Land area: 5ha Constructed in 1999

深圳市华侨城中西部城市综合区城市景观设计
1996年，用地面积85hm²
Landscape Design of Complex Area of
Central and West Area of OCT, Shenzhen
1996 Land area: 85ha

特别致谢
Special Acknowledgment

索引

250 2011年客户名录

252 2011年员工名录

Index

250 **2011 Client List**

252 **2011 Staff List**

2011年客户名录
2011 Client List （按首字母索引）

阿里巴巴（中国）有限公司
Alibaba(China) Co.,Ltd.

百度国际科技（深圳）有限公司
Baidu Internatial Technology (Shenzhen) Co.,Ltd.

北京投资建设集团下属北投贵州城投集团
Beijing Investment & Development Group (Guiyang)

东海航空有限公司
Shenzhen Donghai Airlines Co., Ltd.

佛山市南海区国土城建和水务局
Nanhai Municipal Bureau of Land & Water Resources

福州房地产发展有限责任公司
Fuzhou Real Estate Development Co.,Ltd.

福州市城乡建设发展总公司
Fuzhou Urban and Rural Construction and Development Corporation

广东利海集团有限公司
L'SEA GROUP

广州铁路（集团）公司
Guangzhou Railway (Group) Corporation

贵阳市城乡规划局
Guiyang Municipal Bureau of Urban and Rural Planning

国银金融租赁有限公司
CDB Leasing Co.,Ltd.

海南诗波特投资有限公司
Hainan Sports Investment Co.,Ltd.

海南万科房地产有限公司
Hainan Vanke Real Estate Co.,Ltd.

杭州经济技术开发区城市建设发展中心
Urban Construction and Development Center of Hangzhou Economic and Technological Development Zone

杭州荣兴置业有限公司
Hangzhou Rongxing Real Estate Co.,Ltd.

杭州余杭经济开发建设有限公司
Hangzhou Yuhang Economic Development Co.,Ltd.

河源市深业地产有限公司
Shum Yip (Heyuan) Real Estate Co., Ltd.

深圳华侨城房地产有限公司
Shenzhen OCT Properties Co.,Ltd.

江门市城乡规划局
Jiangmen Municipal Bureau of Urban and Rural Planning

茂名市新宇投资有限公司
Maoming Xinyu Investment Co.,Ltd.

民生金融租赁股份有限公司
Mingsheng Financial Leasing Co.,Ltd

青岛高科产业发展有限公司
Qindao High-Tec Industry Development Co.,Ltd.

清华大学深圳研究生院
Graduate School of Tsinghua University at Shenzhen

山东祥泰森林河湾置业有限公司
Shandong Xiangtai Senlin Hewan Real Estate Co.,Ltd.

山东祥泰颐华置业有限公司
Shandong Xiangtai Yihua Real Estate Co.,Ltd.

深圳莱蒙投资控股有限公司
Shenzhen Top Spring Investment Co.,Ltd.

深圳利通控股有限公司
Shenzhen Leton Holdings Co.,Ltd.

深圳市城市建设开发（集团）公司
Shenzhen Expander (Group) Co.,Ltd

深圳市城市设计促进中心
Shenzhen Center for Design

深圳市地铁集团有限公司
Shenzhen Metro Group Co., Ltd.

深圳市规划和国土资源委员会
Urban Planning, Land and Resources Commission of Shenzhen Municipality

深圳市国际招标有限公司
Shenzhen International Tendering Co.,Ltd.

深圳市海德伦工程咨询有限公司
Shenzhen Hideline Engineering Consultant Co.,Ltd.

深圳市华联置业集团有限公司
Shenzhen Hualian Real Estate (Group) Co.,Ltd.

深圳市华盛房地产开发有限公司
Shenzhen Warmsun Real Estate Development Co.,Ltd.

深圳市科技工贸和信息化委员会
Science, Industry, Trade and Information Technology Commission of Shenzhen Municipality

深圳市明泰润投资发展有限公司
Shenzhen Mingtairun Investment Development Co.,Ltd.

深圳市南山区建筑工务局
Construction Works Bureau of Nanshan District, Shenzhen

深圳市特发集团有限公司
Shenzhen Special Economic Zone Development Group Co.,Ltd.

深圳市投资控股有限公司
Shenzhen Investment Holdings Co.,Ltd.

深圳市祥盛房地产开发有限公司
Shenzhen Xiangsheng Real Estate Development Co.,Ltd.

深圳市新天时代投资有限公司
Shenzhen New Century Investment Co.,Ltd.

深圳市一和雅韵建筑咨询有限公司
Ehow R&D Center

深圳市玉建房地产开发有限公司
Shenzhen Yujian Real Estate Development Co.,Ltd.

深圳盐田港集团有限公司
Shenzhen Yantian Port Group Co.,Ltd.

深圳职业技术学院
Shenzhen Polytechnic

苏州太湖国家旅游度假区管委会
Management Commission of Taihu National Tourism Zone, Suzhou

唐山湾国际游艇发展有限公司
Tangshanwan International Yachts Development Co.,Ltd.

铁道部工程设计鉴定中心
Engineering Design Appraisal Center of the Ministry of Railways

无锡市轨道交通规划建设领导小组（指挥部）办公室
Railway Traffic Planning & Construction Headquater of Wuxi Municipality

仙桃市规划局
Xiantao Municipal Bureau of Planning

仙桃市体育局
Xiantao Municipal Bureau of Sports

襄阳市城乡规划局
Xiangyang Municipal Bureau of Urban and Rural Planning

云南堃驰房地产有限公司
Yunnan Kunchi Real Estate Co.,Ltd

运泰建业置业（深圳）有限公司
Wan Thai Development Limited

中航地产股份有限公司
Avic Real Estate Holding Co.,Ltd.

中山深城建房地产有限公司
Zhongshan Real Estate Development Co.,Ltd of Shenzhen Urban Expander Group

中山市金源房地产开发有限公司
Zhongshan Jinyuan Real Estate Development Co.,Ltd.

中天城投集团股份有限公司
Zhongtian Urban Development Group Co.,Ltd.

中天城投集团贵阳房地产开发有限公司
Guiyang Real Estate Development Co.,Ltd of Zhongtian Urban Development Group -Zhongtian Urban Development Group

中天城投集团贵阳国际会议展览中心有限公司
Guiyang International Conference & Exposition Center Co.,Ltd.

中铁南方投资发展有限公司
China Railway South Investment & Development Co.,Ltd.

2011年员工名录
2011 Staff List

董事长、首席设计师：冯越强　Feng Yves Yueqiang
董事总经理：何　伟　He Wei
董事设计总监：Michel PERISSE / Christophe GAUDIER
设计、技术、运营董事：白宇西　Bai Yuxi
　　　　　　　　　　杨光伟　Yang Guangwei
　　　　　　　　　　沙　军　Sha Jun
　　　　　　　　　　丁　荣　Ding Rong
　　　　　　　　　　叶林青　Ye Linqing
　　　　　　　　　　林建军　Lin Jianjun

陈 虎	Chen Hu	胡海萍	Hu Haiping
陈 然	Chen Ran	胡红凤	Hu Hongfeng
陈 勇	Chen Yong	黄 刚	Huang Gang
陈 凯	Chen Kai	黄 敏	Huang Min
陈成发	Chen Chengfa	黄 煜	Huang Yu
陈春凡	Chen Chunfan	黄 媛	Huang yuan
陈福余	Chen Fuyu	黄邦耿	Huang Banggeng
陈铭恒	Chen Mingheng	黄俊杰	Huang Junjie
陈廷好	Chen Tinghao	黄煦原	Huang Xuyuan
陈小哲	Chen Xiaozhe	黄云涛	Huang Yuntao
陈志兵	Chen Zhibing	黄泽鹏	Huang Zepeng
陈致威	Chen Zhiwei	黄志唐	Huang Zhitang
成 龙	Cheng Long	冀 南	Ji Nan
程婧幂	Cheng Jingmi	简恩洁	Jian Enjie
崔圣美	Cui Shengmei	江志成	Jiang Zhicheng
董孜孜	Dong Zizi	姜 静	Jiang Jing
段安安	Duan An'an	蒋伟军	Jiang Weijun
段志强	Duan Zhiqiang	焦其新	Jiao Qixin
范好仕	Fan Haoshi	金运丰	Jin Yunfeng
方 翀	Fang Chong	景守军	Jing Shoujun
封 颖	Feng Ying	康 娴	Kang Xian
冯 明	Feng Ming	柯文德	Ke Wende
付群霞	Fu Qunxia	邝英武	Kuang Yingwu
傅进毅	Fu Jinyi	雷 霆	Lei Ting
高笑君	Gao Xiaojun	黎云琪	Li Yunqi
龚 侃	Gong Kan	李 貌	Li Mao
龚 婷	Gong Ting	李 密	Li Mi
龚沁华	Gong Qinhua	李 如	Li Ru
顾 康	Gu Kang	李 瑞	Li Rui
郭怀坤	Guo Huaikun	李 洋	Li Yang
何 超	He Chao	李翠华	Li Cuihua
何 健	He Jian	李德嘉	Li Dejia
何 明	He Ming	李富源	Li Fuyuan
何东婉	He Dongwan	李红芳	Li Hongfang
何红令	He Hongling	李开伦	Li Kailun
何荣山	He Rongshan	李艳明	Li Yanming
何同蕾	He Tonglei	梁 旭	Liang Xu
胡 容	Hu Rong	廖晓华	Liao Xiaohua

林 飞	Lin Fei	石丹青	Shi Danqing	肖 雪	Xiao Xue	张海迪	Zhang Haidi
林 富	Lin Fu	宋 俊	Song Jun	肖鸿儒	Xiao Hongru	张厚珺	Zhang Houfei
林超伟	Lin Chaowei	苏 婷	Su Ting	谢 军	Xie Jun	张妙英	Zhang Miaoying
林法伟	Lin Fawei	苏 义	Su Yi	谢永强	Xie Yongqiang	张明珠	Zhang Mingzhu
林文波	Lin Wenbo	苏光军	Su Guangjun	辛佩龙	Xin Peilong	张薇旭	Zhang Weixu
林先文	Lin Xianwen	孙 晨	Sun Chen	熊 亮	Xiong Liang	张英良	Zhang Yingliang
凌立信	Ling Lixin	孙钦学	Sun Qinxue	徐 娜	Xu Na	张志辉	Zhang Zhihui
刘 婧	Liu Jing	孙婷婷	Sun Tingting	徐 宁	Xu Ning	赵 峰	Zhao Feng
刘 猛	Liu Meng	谭 莹	Tan Ying	徐春燕	Xu Chunyan	赵敬信	Zhao Jingxin
刘 宁	Liu Ning	田 静	Tian Jing	徐文娟	Xu Wenjuan	赵灵灵	Zhao Lingling
刘 威	Liu Wei	田金秀	Tian Jinxiu	许 波	Xu Bo	赵雪坤	Zhao Xuekun
刘 渝	Liu Yu	涂 靖	Tu Jing	许 璇	Xu Xuan	郑 竹	Zheng Zhu
刘长寿	Liu Changshou	王 石	Wang Shi	薛 亮	Xue Liang	郑明华	Zheng Minghua
刘国臣	Liu Guochen	王 甜	Wang Tian	薛 山	Xue Shan	郑清容	Zheng Qingrong
刘敏婕	Liu Minjie	王丹青	Wang Danqing	严 浩	Yan Hao	钟复潭	Zhong Futan
刘琴霞	Liu Qinxia	王和文	Wang Hewen	阳 华	Yang Hua	钟洁玲	Zhong Jieling
刘晓鹏	Liu Xiaopeng	王奂为	Wang Huanwei	杨 安	Yang An	钟子贤	Zhong Zixian
刘一锟	Liu Yikun	王漠伦	Wang Molun	杨 玲	Yang Ling	周 洁	Zhou Jie
刘远霞	Liu Yuanxia	王瑞芬	Wang Ruifen	杨 明	Yang Ming	周 茜	Zhou Qian
卢东晴	Lu Dongqing	王莎莎	Wang Shasha	杨琳琳	Yang Linlin	周 熔	Zhou Rong
陆 兵	Lu Bing	王诗峰	Wang Shifeng	杨廷帼	Yang Tingguo	朱 浩	Zhu Hao
罗 洁	Luo Jie	王锡亮	Wang Xiliang	杨晓霞	Yang Xiaoxia	朱理国	Zhu Liguo
罗小萍	Luo Xiaoping	王一军	Wang Yijun	叶 星	Ye Xing	祝 捷	Zhu Jie
毛 婧	Mao Jing	王永峰	Wang Yongfeng	易 侃	Yi Kan	邹国强	Zou Guoqiang
毛飞霞	Mao Feixia	王子鹏	Wang Zipeng	雍兴燕	Yong Xingyan	邹积成	Zou Jicheng
毛同祥	Mao Tongxiang	魏 婷	Wei Ting	袁 冲	Yuan Chong	邹龙清	Zou Longqing
孟 盈	Meng Ying	魏大俊	Wei Dajun	岳连生	Yue Liansheng	邹文斌	Zou Wenbin
倪 昕	Ni Xin	温春苑	Wen Chunyuan	张 斌	Zhang Bin		
聂 凡	Nie Fan	邬 彦	Wu Yan	张 恒	Zhang Heng		
牛玉玮	Niu Yuwei	吴纪元	Wu Jiyuan	张 鲲	Zhang Kun	外籍：	
欧阳霞	Ouyang Xia	吴俊杰	Wu Junjie	张 蕾	Zhang Lei	AUDREY NOEL	
潘志坚	Pan Zhijian	吴晓慧	Wu Xiaohui	张 翔	Zhang Xiang	ETIENNE CHAMPENOIS	
漆文亮	Qi Wenliang	吴英家	Wu Yingjia	张 勇	Zhang Yong	Franck CONSTANS	
丘 蕊	Qiu Rui	吴志刚	Wu Zhigang	张 淼	Zhang Miao	IHEB GUERMAZI	
邱 敏	Qiu Min	夏 淼	Xia Miao	张 平	Zhang Ping	JOANA ITURRIA	
阮慈惠	Ruan Cihui	向 娇	Xiang Jiao	张昌蓉	Zhang Changrong	MARIAROSARIA GIULIANO	
沈懋文	Shen Maowen	向 凯	Xiang Kai	张光理	Zhang Guangli	TIEME WILLEMS	
盛洁菲	Sheng Jiefei	肖 浪	Xiao Lang			VIRGILE CANNAVO	

实习生

蔡 颖	Cai Ying
陈苗苗	Chen Miaomiao
陈世圣	Chen Shisheng
杜 肖	Du Xiao
付迎霞	Fu Yingxia
桂 超	Gui Chao
胡芳珍	Hu Fangzhen
胡宇超	Hu Yuchao
黄晓仪	Huang Xiaoyi
季路尧	Ji Luyao
荆宇璐	Jing Yulu
李志祥	Li Zhixiang
林猛东	Lin Mengdong
林伟建	Lin Weijian
林志聪	Lin Zhicong
刘宝森	Liu Baosen
刘诛雯	Liu Shuwen
秦荣明	Qin Rongming
苏婧婷	Su Jingting
涂晓蕾	Tu Xiaolei
温绍力	Wen Shaoli
邬 睿	Wu Rui
吴必超	Wu Bichao
张乃昌	Zhang Naichang
章 颜	Zhang Yan
祝艺苗	Zhu Yimiao

特别感谢给予Aube(欧博设计)信任的各界朋友、客户，感谢所有全心投入的欧博人

Grateful thanks to all our friends, colleagues and clients for your trust and surport, as well as to our devoted Auber

主编寄语
EDITORIAL'S LETTER

胸怀·远见

在去年的年会上，面对着浩瀚的海洋，我曾与欧博人分享的感悟是《梦想·现实》；今年，面对着鳞次栉比的钢铁森林，愿与各位共勉的主题是《胸怀·远见》。当梦想照进现实，我们怀着豪情坚定地迈过了2011年。过去的一年已成为历史，但是希望我们能够"以史为鉴"，共同感悟2011年的付出与收获。相信今天的这份思考必将会成为今后我们生活和工作中的宝藏！

锐气藏于胸

纵观古今中外，大凡有作为者，除了拥有卓越的才智及执著追求的精神之外，他们还有一个共性——胸怀大度的精神。《史记•魏公子列传》中有这样一个典故：信陵君为人仁爱宽厚礼贤下士，无论才能大小，他都谦恭有礼地同他们交往，从来不敢因自己富贵而轻慢士人。他听说大梁城东门的守门人侯嬴很有才能，于是去拜请他。侯嬴毫不谦让地直接坐在公子空出的尊贵座位，让信陵君赶车。车到闹市后，侯嬴故意与好友朱亥长时间交谈，将信陵君晾在一边，但公子的面色始终和悦谦恭。回到公子府后，信陵君大宴宾客，推侯嬴坐在上席，并亲自给他敬酒。这则小事反映出信陵君博大的胸怀，正是因为他有如此开阔的胸襟、容人的气度，才能使能人贤士聚之门下，并为"窃符救赵"立下奇功，成就了信陵君战国四公子的美谈。

常人的胸怀又该如何呢？联想到当下的我们，在生活中、工作中不如意之事比比皆是。当苦闷、伤痛和被人误解等烦恼像潮水般涌来的时候，我们能否做到胸怀坦荡？大家不妨静下心来，想一想、问一问：我们大家一起共事的过程中

——观点不统一时；
——工作方法不同时；
——性格差异显现时；
——能力高低迥然时…

作为一名组织者或领导者，应该以什么胸怀对待？
——是先否定对方的观点，还是先听一听其见解？
——是先讨论一下可依据并行之有效的方法，还是先认定唯我正确？
——是先有"性格不合便无法共事"的态度，还是先采取以情动人、求同存异、以共同目标为大局的合作之途？
——是先以"恶其余胥"式的心态全盘否定一个人能力，还是先运用"用其所长、避其所短"的爱才方式去寻求共同进步？

时刻拥有博大的胸怀，是我们共同追求的理想状态；即使不能做到每时每刻，我们也要有时刻积极尝试的觉悟。只有放弃了烦恼、怨恨和不平，才能得到友情、快乐和幸福；只有舍掉了狭隘、偏激或毫无意义的计较，才能换来祥和、宁静、畅快融洽的关系。那么这时候的我们，就一定会拥有磊落坦荡、无私无畏和志存高远的品格！

远见行于事

事实上，拥有了博大胸怀的人，也就同时拥有了高瞻远瞩的见识。如同登山，如果你的目标是绝顶，就不会在小峰流连忘返；如果你的目标是大海，就不会在小溪边得意忘形。一旦拥有远见，就能站得高看得远。而这个目标的达成，则是建立在脚踏实地、锲而不舍、时时进步、朝着既定目标奋勇前进的基础上。"不积跬步，无以至千里"，胸怀远见的人，不会为一时的名利影响前行的道路，不会为别人的一句顶撞、一个误解或一时诽谤而耿耿于怀。他们所做的，是从大局出发，从而把握全局并且争取判断局势走向，同时他们也在不断超越自己——这就是所谓的虚怀若谷、志存高远、远见卓识。

"锐气藏于胸，和气浮于面。"心胸有多大，事业的进步、成就和认可度就有多大；心胸有多广，未来的发展潜力和境界就会有多宽广。愿大家都能在新的一年里，心怀包容、胸怀大"智"；也愿阳光和幸福与你同在。

"毋意，毋必，毋固，毋我。"
希望《论语》中的箴言，成为每个欧博人在2012年工作中互勉的目标！

欧博设计
董事长、首席设计师 冯越强
2012年元月18日于深圳威斯汀酒店

TOLERANCE • VISION

In the annual meeting last year, we shared dream and reality in the face of a vast sea. This year, surrounded by Shenzhen's steel forest, we will focus on the theme of tolerance and vision. Bearing dreams in mind, we spent the year of 2011 with passion and determination. Though the past year has become history, we hope to "learn from history" and reflect on what we have paid and gained in 2011. It is believed that the reflection today will surely turn into treasure of our future work and life.

Ambition Hidden in Heart

At all times and in all countries, most of the successful people have something in common----a broad mind as well as superior talent and persistence. In Biographies of Lord Wei in Records of the Grand Historian there is a classical allusion that goes like this. As a benevolent and generous man, Lord Xinling treated men of talent with courtesy, whatever their talents were and never disrespected the literati for the sake of his status and fortune. Having heard that Hou Ying, the gatekeeper of the east gate of Daliang, was talented, he visited him and invited him to be his company. Hou Ying sat directly on the seat of honor left by the lord without modesty, while Lord Xinling was left to drive the cart. When they reached the downtown, Hou Ying talked to his good friend Zhu Hai intentionally for a long time, neglecting Lord Xinling who was still as kind and humble as he could be. After they got to the mansion of Lord Xinling, the Lord held a grand banquet in Hou Ying's name, recommended him to sit on a seat of honor and proposed a toast to him, which shows the broad mind of Lord Xinling. It is exactly due to this quality that talented people gather around him and great success is achieved in "stealing the commander's tally to save states of Zhao" which contributes to his status as one of the most respectable Four Lords of the Warring States.

What should ordinary people be like? Things do not always happen as we expect in life and work. When we are overwhelmed with anguish, pains, misunderstandings, and other troubles, can we be broad-minded? We might as well calm down, think about it and ask ourselves: in our daily work

——When opinions are not unified,
——When working methods are different,
——When differences in personality appear,
——When remarkable variance is shown in ability,
what should we do as an organizer or a leader?
——Should we deny others' views directly, or listen to others' views first?
——Should we have a discussion and figure out an effective method with others or assert the correctness of our own opinion?

Should we hold the attitude that "people of incompatible personalities cannot work together" in advance, or bear goals in mind and put aside minor differences so as to seek common ground for cooperation?

Should we totally negate a person's ability "just because he has shortcomings" or should we give credit to his strong points and avoid his weak points so as to seek common progress?

The ideal state that we all pursue is to be broad-minded at all times. Even if we cannot keep that state at every moment, we should try actively at all times. Only when we abandon annoyance, resentment and grievance can we attain friendship, joy and happiness. Only when we let go of narrowness, bias and meaningless arguments can we cultivate such qualities as integrity, selflessness, fearlessness and great ambition.

Vision in actions

In fact, people who are broad-minded are also people with vision. If your goal is the highest summit, you will not linger on small peaks. If your goal is the sea, you will not become conceited just standing beside brooks. Once you have foresight, you can stand higher and see further. However, the achievement of this goal is based on step-by-step, perseverant and progressive efforts to move on. "A journey of thousands of miles may not be achieved without accumulation of every single step." Broad-minded people will not stop their steps because of temporary fame and gain, nor will they take it personally when others contradict, misunderstand or slander them. What they do is start from the overall situation, grasp it and strive to judge its trend while surpassing themselves constantly. This is what is called modesty, ambition and vision.

"Hide ambition and show kindness". The broader your mind is, the more successful your career will be. The broader your mind is, the greater your future potential of development will be. May everyone be broad-minded and keep "wisdom" in heart in the coming year! May sunshine and happiness always be with you!

"No foregone conclusions, no arbitrary predeterminations, no obstinacy, and no egoism."

May the maxim in Confucius become a common goal for everyone in AUBE in 2012!

AUBE Conception
President & Chief Designer: Yves Yueqiang Feng
The Westin Hotel, Shenzhen
January 18, 2012

欧博设计 法国欧博建筑与城市规划设计公司
深圳市欧博工程设计顾问有限公司

欧博设计网址: www.aube-archi.com
中国深圳 华侨城生态广场C栋二层 / 华侨城生态广场A栋201
电话:(86 755)26930794 26930795
传真:(86 755)26918376
法国巴黎 1,RUE PRIMATICE 75013 PARIS FRANCE
TEL:(33.1) 45709200
FAX:(33.1) 45709819

Chief Editor: FENG Yves Yueqiang
Executive Editor: GAO Xiaojun
Advisory Editors: BAI Yuxi   YANG Guangwei   Sha Jun   Lin Janjun   Michel PERISSE
Editors: Jiang Jing
Translators: Zhang Yan   HE Tonglei   Cheng Jingmi

主　　编: 冯越强
执行主编: 高笑君
顾问编委: 白宇西　杨光伟　沙　军　林建军　Michel Perisse
编　　委: 姜　静
翻　　译: 章　彦　何同蕾　程婧幂

图书在版编目(CIP)数据

2011欧博设计／法国欧博建筑与城市规划设计公司，深圳市欧博工程设计顾问有限公司编著. —北京：中国建筑工业出版社，2012.3
ISBN 978-7-112-14087-9

I.①2··· II.①法··· ②深··· III.①建筑设计—作品集—中国—现代 IV.①TU206

中国版本图书馆CIP数据核字（2012）第032064号

责任编辑：常 燕 袁瑞云

2011欧博设计

法国欧博建筑与城市规划设计公司

深圳市欧博工程设计顾问有限公司 编著

*

中国建筑工业出版社出版、发行（北京西郊百万庄）
各地新华书店、建筑书店经销
AUBE 欧博设计制版
深圳市国际彩印有限公司印刷

*

开本：787×1092毫米 1/12 印张：22 字数：536千字
2012年3月第一版 2012年3月第一次印刷
定价：238.00元
ISBN 978-7-112-14087-9
(22133)

版权所有 翻印必究
如有印装质量问题，可寄本社退换
（邮政编码 100037）